Statistical Physics of Crystal Growth

Yukio Saito

Department of Physics, Keio University
Yokohama, Japan

Statistical Physics of Crystal Growth

World Scientific
Singapore • New Jersey • London • Hong Kong

Published by

World Scientific Publishing Co. Pte. Ltd.

P O Box 128, Farrer Road, Singapore 912805

USA office: Suite 1B, 1060 Main Street, River Edge, NJ 07661

UK office: 57 Shelton Street, Covent Garden, London WC2H 9HE

Library of Congress Cataloging-in-Publication Data
Saito, Yukio, 1948–
    Statistical physics of crystal growth / Yukio Saito.
       p.   cm.
    Includes bibliographical references and index.
    ISBN 9810228341
    1. Crystal growth. 2. Statistical physics. I. Title.
QD921.S23   1996
548'.5--dc20   96-27446
                                CIP

British Library Cataloguing-in-Publication Data
A catalogue record for this book is available from the British Library.

First published 1996
First reprint 1998

Printed in Singapore.

Preface

The study of crystal and its growth has a long tradition. The art of crystal growth is now well developed to manufacture various high quality crystals for electronics, but its scientific understanding is still developing. The modern technologies and observation methods of the atomic scale make possible and also require the microscopic understanding of the growth mechanisms to control the quality of the product.

Since the crystal growth is a typical example of a system far from equilibrium which involves the first-order phase transition, its comprehension involves many disciplines. In these two decades novel theoretical contributions are given from the field of statistical physics: Kosterlitz-Thouless phase transition of the surface roughness in equilibrium surface structure, and the microscopic solvability criterion for the selection of the dendritic crystal growth.

This note tries to give the concise and precise overview on the fundamental scientific aspects of crystal growth from the simplest phenomenological argument to the latest discoveries. Technology and art of crystal growth are omitted in this book. Important coupling of the hydrodynamics to the crystal growth are beyond the capability of the author and are not touched upon.

The note is based on my introductory lectures on the fundamental and physicomathematical aspects of the crystal growth given at a winter school of Japanese Association of Crystal Growth and in graduate courses in many universities in Japan; Keio, Ochanomizu, Hokkaido, Kyoto and so on.

I am especially grateful to H. Müller-Krumbhaar. From discussions and a longlasting collaboration with him, I learned most of the fundamental theories of crystal growth described in this book. The collaborations and enlightening discussions with C. Misbah, E. Brener, D. Temkin and M. Uwaha are also deeply acknowledged. To M. Uwaha I am thankful for his reading through and giving comments on the manuscript. I am also indebted to T.Ohta, K.Wada, M.Kitamura, who have invited me to their universities and given chances to teach the courses which eventually resulted in this book. Lastly I dedicate my heartful appreciation to late Prof. R. Kubo, who introduced me to the fascinating world of statistical physics.

Yukio Saito

Contents

1 Introduction

We are living in the age of informatics; In computers a mass of information is handled in high speed by integrated circuits built on a semiconductor crystal. By means of a laser light emitted from semiconductor crystals, information written in a compact disc at high density is read out. Through the optical fiber runs information as the optical signal with high density and in the extremity of speed. In this information technology, semiconductor crystals with high quality are required. There are also many other crystalline materials utilized in our life; steal for cars, quartz in watch, sugar and salt and so on. To fabricate crystals under our control we have to know the mechanism of crystal growth.

There are also various crystals grown in nature; minerals, gems as diamonds, snow etc. They contain information about the history how and the environment in which they are grown: Snow is called a letter from the sky, diamond a letter from deep in the earth. In order to understand their messages we have to know the dynamics of crystal growth.

Crystal has an ordered arrangement of atoms or molecules in microscopic scale. The microscopic regularity shows up in the symmetry as is observed in various diffraction patterns. Also the regular arrangement of atoms brings about the symmetry in crystalline shape: Some crystals take simple forms as polyhedra as shown in Fig.1.1, reflecting their symmetry. Some crystals have complicated forms as dendrite, as shown in Fig.1.2 and in Fig.1.3. They are complex in the sense that all the snow flakes, for example, look similar with six arms but none of them are completely the same. We want to know how the crystal shapes are determined. The difference in crystal shape should be brought about by the difference in the controlling mechanisms of the growth dynamics. The relation between the growth mechanism and the resulting growth morphology is to be explored.

There are many textbooks and monographs on crystal growth [74, 48, 120, 85]. I intend to give in this book a concise but precise overview on the fundamental theories on crystal growth from the viewpoint of statistical physics, especially on the recent developments in the pattern formation in the diffusion field. In Part I, the ideal growth formulae are derived from a thermodynamical point of view. "Ideal" here means that all the thermodynamic driving force for the phase transition is invested for the crystal growth. The growth formula thus gives the maximum growth velocity. In reality there are many hindrances against the growth. The largest effect takes place at the crystal surface, where the growth process takes place. In Part II, the equilibrium structure of the crystal surface is studied microscopically, and the surface phase transition of roughening is discussed. For an atomically smooth surface, surface kinetics governs the crystal growth and the growth laws in this situation is discussed in Part III. For an atomically rough surface, material transport or the heat conduction controls the crystal growth. In this case, macroscopic morphology of the crystal surface is strongly influenced, and complex pattern formation is induced. The topic will be discussed in Part IV.

(a)   (b)

Figure 1.1: Polyhedral crystals of (a) NaCl [79] and (b) quartz. (Courtesy by I.Sunagawa).

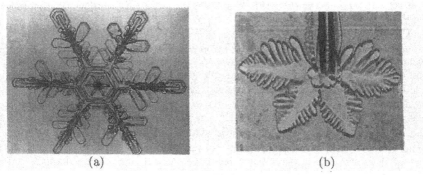

(a)   (b)

Figure 1.2: Dendritic crystals of (a) snow and (b) ice. (Courtesy by Y.Furukawa).

(a)   (b)

Figure 1.3: Irregular dendrites of (a) MnO_2 (Courtesy by N.Osada) and (b) Au on Pt(111) [41].

Part I
Ideal Growth Laws

Crystal growth is an example of a dynamical first order phase transition. A stable phase, crystal, grows out from a metastable phase, melt or vapor. The driving force for the growth is the chemical potential difference of the stable and the metastable phases. The simple assumption of the linear response that the growth velocity is proportional to the driving force gives an ideal linear growth laws. When the crystal is finite, or when the interface deforms, the surface tension has to be included in the thermodynamics of the crystal growth. Surface tension plays an important role in the determination of the shape of a finite sized crystal, or the deformation of the flat interface.

2 Melting and Solidification: First-Order Phase Transition

At low temperatures almost all the materials order in crystalline form where atoms are arranged regularly. (Fig.2.1a). This is the configuration with the minimum interaction energy E of atoms or molecules. At a high temperature atoms break the regular structure since it is less free and with little entropy. In an irregular arrangement as in liquid (Fig.2.1b) and in gas (Fig.2.1c), molecules move more freely and thus gain the entropy in spite of the energy cost. For a given temperature T and pressure p, the equilibrium state is determined by the second law of thermodynamics

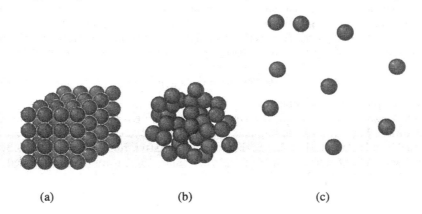

(a)   (b)   (c)

Figure 2.1: Arrangement of atoms (a) in a crystal, (b) in a liquid and (c) in a gas phase.

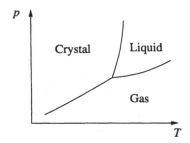

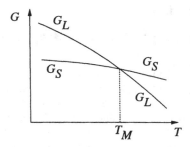

Figure 2.2: Equilibrium phase diagram in the temperature T and the pressure p phase space.

Figure 2.3: Isobaric variation of the Gibbs free energies of the liquid, G_L, and of the crystal G_S as functions of temperature.

[121] such that the Gibbs free energy $G = E - TS + pV$ is minimized. Here V is the volume of the system. At low temperatures with small values of T the entropy S contributes less than the enthalpy $H = E + pV$ and the crystal with low H is realized (Fig.2.2). At high temperatures, on the other hand, the entropy S gives the dominant contribution and the configuration with a large S has the low free energy G: the liquid or gas phase is realized. Two phases can coexist on a coexistence curve.

We consider in the following a liquid-crystal phase transition as an example. Isobaric variation of the Gibbs free energies of liquid and crystal, G_L and G_S respectively, are shown in Fig.2.3 as functions of temperature T. At a low temperature G_L lies higher than G_S showing that the crystal is stabler than the liquid, but on increasing the temperature, G_L decreases faster than G_S due to the large entropy S_L of liquid compared to S_S of crystal. At a melting point $T_M(p)$, G_L and G_S cross with each other

$$G_S(T_M, p) = G_L(T_M, p), \qquad (2.1)$$

and the liquid becomes stabler for $T > T_M$.

On heating the crystal under a constant pressure, its temperature first increases, as shown in Fig.2.4. At a melting point T_M the temperature stops increasing and the applied heat is consumed to change the state of the matter from crystal to liquid. Since the absorbed heat does not appear explicitly as a temperature rise, it is called the latent heat L. The first law of thermodynamics says that the applied heat changes into the work pdV done to the environment and the increment of the internal energy dE. Since the pressure is kept constant, the heat changes into the enthalpy $H = E + pV$: $dH = dE + d(pV) = dE + pdV$. Thus the latent heat observed at a melting point corresponds to the enthalpy difference of the two phases as $L = H_L(T_M, p) - H_S(T_M, p)$, with H_S and H_L being the enthalpy of the crystal and liquid phases respectively.

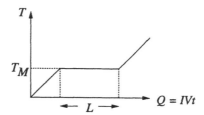

Figure 2.4: Isobaric variation of the temperature due to the heating, for example, by an electric current I under the voltage V. At the melting temperature T_{M}, crystal starts to melt, and until the completion of melting, system absorbs the latent heat L.

Since Gibbs free energies of both phases are equal at the melting point, $G_{\mathrm{S}} = H_{\mathrm{S}} - T_{\mathrm{M}}S_{\mathrm{S}} = G_{\mathrm{L}} = H_{\mathrm{L}} - T_{\mathrm{M}}S_{\mathrm{L}}$, the latent heat is proportional to the entropy difference $\Delta S = S_{\mathrm{L}} - S_{\mathrm{S}}$:

$$L = H_{\mathrm{L}} - H_{\mathrm{S}} = T_{\mathrm{M}}\Delta S. \qquad (2.2)$$

According to thermodynamics [121], entropy S is the temperature derivative of the Gibbs free energy as $S = -(\partial G/\partial T)_p$. Thus the phase transition with a latent heat is associated with a discontinuity in the slope of the Gibbs free energy

$$\left(\frac{\partial G_{\mathrm{L}}}{\partial T}\right)_p - \left(\frac{\partial G_{\mathrm{S}}}{\partial T}\right)_p = -\frac{L}{T_{\mathrm{M}}}. \qquad (2.3)$$

Ehrenfest named this type of phase transition with a discontinuity in the first derivative of G as a first-order phase transition .

3   Crystal Growth from the Melt

Now let us cool the liquid to a temperature T below the melting point T_{M}. The Gibbs free energy of a crystal $G_{\mathrm{S}}(T,p)$ is lower than that of a liquid $G_{\mathrm{L}}(T,p)$ as shown in Fig.2.3, and the true equilibrium state is a crystalline phase. However, the whole liquid cannot instantaneously turn into the crystal. We often experience that the liquid is supercooled for a long while near the melting point. Eventually, a small crystalline nucleus is formed in the liquid, and then it grows. The crystal grows by the advancement of a crystallization front. The evolution is driven by the second law of thermodynamics so as to minimize the Gibbs free energy at a given temperature and pressure [121]. The driving force of the crystal growth is the difference of the Gibbs free energies of the liquid and of the crystal phases: $\Delta G(T,p) = G_{\mathrm{L}}(T,p) - G_{\mathrm{S}}(T,p)$. Since it vanishes at the melting point T_{M}, one can expand ΔG up to the first order of the undercooling $\Delta T = T_{\mathrm{M}} - T$ as

$$\Delta G \approx \left(\frac{\partial G_{\mathrm{L}}}{\partial T}\right)_p (T - T_{\mathrm{M}}) - \left(\frac{\partial G_{\mathrm{S}}}{\partial T}\right)_p (T - T_{\mathrm{M}}) = L\frac{\Delta T}{T_{\mathrm{M}}}. \qquad (3.1)$$

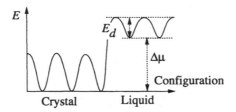

Figure 3.1: Schematics of the potential surface in configuration space. Crystal phase corresponds a stable phase, liquid a metastable phase, and in between is the diffusion activation energy.

In the last equality we used the relation (2.3).

Under this chemical driving, the crystal grows. For a liquid molecule to be incorporated into the crystalline order, it has to change the configuration. But around the liquid molecule there is a high density of other molecules and they hinder the free motion of the molecule. It can mainly vibrate around its average position with a frequency ν, which is of the order of that of the lattice vibration. In order for a liquid molecule to change the configuration drastically, it has to overcome the energy barrier E_d of the molecular diffusion, as is shown in Fig.3.1. At a temperature T a molecule acquires the energy fluctuation E_d with a probability proportional to the Boltzmann weight $\exp(-E_d/k_BT)$, where k_B is the Boltzmann constant. Therefore, the crystallization rate per unit time is given as $\nu \exp(-E_d/k_BT)$. There is, however, a counter effect, namely the melting of a crystal molecule into the liquid state. Since the Gibbs free energy per molecule, called chemical potential $\mu = G/N$, is higher in the liquid phase than in the crystal phase, $\mu_L > \mu_S$, the rate of melting is smaller than that of crystallization by a factor $\exp(-\Delta\mu/k_BT)$. Here $\Delta\mu$ is the difference of the chemical potentials between the liquid and the crystal phases: $\Delta\mu = \mu_L(T,p) - \mu_S(T,p)$. By the crystallization of one molecule, the solidification front increases its height by a molecular height a. Thus the growth rate is given as

$$V = a\nu \exp\left(-\frac{E_d}{k_BT}\right)\left[1 - \exp\left(-\frac{\Delta\mu}{k_BT}\right)\right]. \tag{3.2}$$

In terms of the liquid viscosity η or the diffusion constant D one can describe the formula as

$$V = \frac{k_BT}{6\pi a^2\eta}\left[1 - \exp\left(-\frac{\Delta\mu}{k_BT}\right)\right] = K\left[1 - \exp\left(-\frac{\Delta\mu}{k_BT}\right)\right]. \tag{3.3}$$

where the Einstein-Stokes relation [56]

$$\nu a^2 \exp\left(-\frac{E_d}{k_BT}\right) = D = \frac{k_BT}{6\pi\eta a} \tag{3.4}$$

is used. Here $K = k_B T/6\pi a^2 \eta$ is called the kinetic coefficient. The linear growth law (3.2) or (3.3) is called the Wilson-Frenkel formula for the melt growth [196, 63], and for the small undercooling ΔT it is approximated as

$$V \approx K \frac{\Delta \mu}{k_B T} \approx K_T \Delta T, \qquad (3.5)$$

where $K_T = Kl/k_B T T_M$ with $l = L/N$ being the latent heat per molecule. At small undercooling ΔT, the growth rate is proportional to the undercooling.

The growth rate thus obtained is an ideal one and valid only when the interface is atomically rough and the whole surface is accessible for the crystallization. In the actual crystal growth the roughness of the interface controls the growth rate. Also the derivation of Eq.(3.2) lacks the consideration of latent heat and of the associated temperature increase near the interface. The effect of heat transport is important in determining the growth rate and the morphology of the crystal, but that will be discussed later in Part III.

The Wilson-Frenkel formula says that the growth velocity V should drop drastically at low temperatures, since the liquid viscosity η increases exponentially. In a molecular dynamics simulation of the crystal growth of the simple molecule system [39], on the other hand, the growth rate is not limited by the mobility of the atoms in the bulk liquid: The kinetic coefficient K in Eq.(3.5) is found to be proportional to the temperature. Explanation is given such that the precursor of the crystalline order is already formed in the liquid phase near the interface, and the collective motion of the liquid facilitates the crystal growth without the activation barrier [137]. Thus, general and microscopic consideration on the kinetic coefficient seems worth to be studied.

4 Vapor Growth

Crystal can grow not only from a melt but also from a gas phase, as shown in Fig.4.1. For instance, the snow is a crystallized water from the vapor (Fig.1.2a), whereas the ice is the one grown from the liquid water (Fig.1.2b). In a semiconductor industry, vapor deposition of a thin film on the substrate is an important technology to fabricate materials with a controlled design and new functions. Molecular beam epitaxy (MBE) or atomic layer epitaxy (ALE) are among these modern technologies.

For a gas phase an ideal gas is a good approximation due to its low density. At a temperature T and a pressure p, the velocity distribution of a monatomic ideal gas follows the well known Maxwell distribution [121]

$$P(\mathbf{v})d\mathbf{v} = \left(\frac{m}{2\pi k_B T}\right)^{3/2} \exp\left(-\frac{m\mathbf{v}^2}{2k_B T}\right) d\mathbf{v}, \qquad (4.1)$$

where $\mathbf{v} = (v_x, v_y, v_z)$ is the velocity of an atom and m is its mass. In a unit time,

$$f = \int_{-\infty}^{\infty} dv_x \int_{-\infty}^{\infty} dv_y \int_{-\infty}^{0} dv_z n|v_z|P(\mathbf{v}) = \frac{p}{\sqrt{2\pi m k_B T}} \qquad (4.2)$$

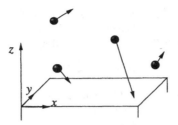

Figure 4.1: Vapor atoms depositing on a flat surface.

number of atoms impinge on a unit crystal surface normal to the z-direction as shown in Fig.4.1, and try to crystallize. Here $n = p/k_BT$ is the average number density of gas atoms.

Inversely, there are some atoms desorbing from the crystal surface at a finite temperature T. The desorption flux is independent of the deposition flux from a gas phase, but is a function of the temperature. If the crystal is in the gas with a saturation pressure, $p_{\text{eq}}(T)$, the deposition rate balances with the desorption rate: The desorption flux from the crystal f_{des} is equal to the deposition flux f_{eq} of a gas at a saturation pressure $p_{\text{eq}}(T)$. Assuming that an atom is cubic with a linear dimension a, the net atomic flux in an atomic area a^2 contributes to the crystal growth, and the atomic height increases by a. The growth rate is thus written as

$$V = a^3(f - f_{\text{eq}}) = \frac{\Omega(p - p_{\text{eq}})}{\sqrt{2\pi m k_B T}}. \tag{4.3}$$

Here Ω is the specific volume of a single molecule $\Omega = a^3$. This linear growth law is called the Hertz-Knudsen's formula [77, 116].

Since the chemical potential of an ideal gas is written as $\mu_{\text{G}}(T, p) = \mu_{\text{G}}(T, p_0) + k_BT \ln(p/p_0)$ [121], the chemical potential difference of the gas and the crystal is written as $\Delta\mu = \mu_{\text{G}}(T, p) - \mu_{\text{S}}(T) = \mu_{\text{G}}(T, p) - \mu_{\text{G}}(T, p_{\text{eq}}(T)) = k_BT \ln(p/p_{\text{eq}})$, or $p = p_{\text{eq}} e^{\Delta\mu/k_BT}$. Thus the Hertz-Knudsen formula is represented as

$$V = \Omega f_{\text{eq}}\left[\exp\left(\frac{\Delta\mu}{k_BT}\right) - 1\right] \equiv K\left[\exp\left(\frac{\Delta\mu}{k_BT}\right) - 1\right] \approx K\frac{\Delta\mu}{k_BT}. \tag{4.4}$$

The growth rate V is proportional to the driving force $\Delta\mu$ for small $\Delta\mu$. The kinetic coefficient K is now obtained as $K = \Omega f_{\text{eq}}$. The relation (4.4) is valid only in the ideal situation of a rough surface with the fast transport of materials in the gas phase. For a smooth and flat surface, the growth law should be modified as will be described in Part III.

5 Solution Growth

It is technically hard to crystallize a material with a high melting temperature, but by solving it into some solution crystallization becomes easier at low temperatures. For example, NaCl melts at 800°C and melt growth has to be performed at a high temperature. If it is solved into hot water, the solution becomes supersaturated by evaporation and a large crystal can be grown even at a room temperature.

In the case of solution growth, there should be a solute atom in front of the interface to crystallize. The probability to find a solute atom at a certain crystallization point with a unit volume a^3 is $a^3 c$ for a solution with a concentration c. This atom is oscillating around the average position with a frequency ν, and tries crystallization. In the solution, however, the solvent molecule makes some chemical bonding with the solute molecules. Thus, for the solute molecules to be incorporated into the crystalline structure, the solvent molecules have to be desolved from the solute. For this desolvation process there is an energy barrier E_{desol}. Among ν trials of solidification, the rate which overcomes the desolvation energy barrier is given by the Boltzmann weight $\exp(-E_{\text{desol}}/k_B T)$. Then the velocity of crystallization is given by $V_{\text{cry}}(c) = a\nu(ca^3)\exp(-E_{\text{desol}}/k_B T)$. There is an inverse process of melting of crystal molecules. Its rate is determined by the temperature, and should balances the crystallization from the solution with the equilibrium concentration c_{eq}: $V_{\text{mel}} = V_{\text{cry}}(c_{\text{eq}})$. The net rate of crystallization is

$$V = V_{\text{cry}}(c) - V_{\text{mel}} = \nu a^4 \exp\left(-\frac{E_{\text{desol}}}{k_B T}\right)(c - c_{\text{eq}}). \tag{5.1}$$

The chemical potential of a dilute solution with a concentration c is expressed as $\mu_{\text{sol}}(T, c) = \mu_{\text{sol}}(T, c_0) + k_B T \ln(c/c_0)$. The chemical potential of the crystal is equal to that of a solution with an equilibrium concentration c_{eq} as $\mu_S(T) = \mu_{\text{sol}}(T, c_{\text{eq}})$. Then the excess chemical potential of the supersaturated solution is written as

$$\Delta\mu = \mu_{\text{sol}}(T, c) - \mu_S(T) = k_B T \ln\frac{c}{c_{\text{eq}}} \approx k_B T\left(\frac{c}{c_{\text{eq}}} - 1\right). \tag{5.2}$$

The growth rate is expressed in terms of the chemical potential difference $\Delta\mu$ as

$$V \equiv K\left[\exp\left(\frac{\Delta\mu}{k_B T}\right) - 1\right] \approx K\frac{\Delta\mu}{k_B T}. \tag{5.3}$$

The kinetic coefficient is now defined as $K = \nu a^4 c_{\text{eq}} \exp(-E_{\text{desol}}/k_B T)$.

The linear growth law (5.2) is valid only in an ideal case when the interface is rough and the whole interface contributes to the growth. Actually by the solidification the number of solute molecules in front of the interface decreases, and to compensate this solute deficiency, molecules have to be transported from the far end of the solution. By including this diffusional material transport, we reach to the different growth law. But the detail will be discussed later in Part VI.

Exercise: If the transport of materials in the ambient phase is important, it introduces a length scale l which characterizes the effective range for the material to be incorporate from the environment. Find in this case the growth formula of an [100]-face of a crystal with a unit cell $a_1 \times a_2 \times a_3$.

Answer: The number of atoms to be incorporated on a unit of [100]-face with an area $a_2 a_3$ is given by $c a_2 a_3 l$ with a solute concentration c. Since the interface advances a height a_1 by the crystallization of one atom, the growth rate is described as

$$V_1 = l\nu\Omega e^{-E_{\text{desol}}/k_B T}(c - c_{\text{eq}}), \tag{5.4}$$

where $\Omega = a_1 a_2 a_3$ is the volume of the unit cell [23].

6 Equilibrium Shape

So far we considered the growth of a crystal with a flat and infinitely large interface. If the interface deforms, the interface area changes and the associated energy change has to be considered. Also in the early stage of crystal growth, it starts from a small embryo. During growth embryo size increases, and the energy cost at the interface changes. Thus the interface energy has effects both on the crystal shape and the growth dynamics. In this section we focus on its effect on the crystal shape in equilibrium [200, 121, 6]. Even though the realization of equilibrium shape is difficult due to its long relaxation time, there are some experiments with very small crystals [78, 79, 136], or with a crystal with fast material transport [147].

Due to the energy cost at the interface, a crystal nucleus with a finite size and shape cannot coexist with a mother phase at a bulk transition point, where the chemical potentials of two phases are equal: $\Delta\mu = 0$. Finite $\Delta\mu$ is necessary to compensate the energy increase at the interface and to keep the small nucleus in the stationary state.

Let us assume that $z = \zeta(x, y)$ describes the phase boundary between the crystal

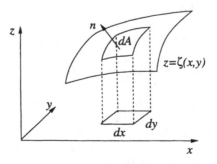

Figure 6.1: Crystal surface.

and liquid phases (Fig.6.1). Then the free energy gain by the solidification of N_S atoms is given as

$$G_b = N_S(\mu_S - \mu_L) = -\frac{V}{\Omega}\Delta\mu = -\frac{\Delta\mu}{\Omega}\int dxdy\zeta(x,y), \qquad (6.1)$$

where Ω and V are the atomic volume and total volume of the crystal cluster, respectively, and the volume integration is taken over the crystal phase ($z < \zeta(x,y)$). The bulk free energy G_b is negative for supercooled melt ($\mu_S < \mu_L$). There is an energy increase due to the formation of an interface. The interface free energy per area $\gamma(p_x, p_y)$ is a function of orientation of the interface $\mathbf{n} = (-p_x, -p_y, 1)/\sqrt{1 + p_x^2 + p_y^2}$ where $p_x \equiv \partial\zeta/\partial x$ and $p_y \equiv \partial\zeta/\partial y$ are the interface gradients. The total surface free energy is written as

$$G_s = \int dA\gamma(p_x, p_y) \equiv \int dxdyf(p_x, p_y). \qquad (6.2)$$

Since the projection of the surface area dA in z-direction is equal to the area $dxdy$ in xy-plane as $n_z dA = dxdy$, the free energy density f per unit area in xy-plane in Eq.(6.2) is given as

$$f(p_x, p_y) = \gamma(p_x, p_y)\sqrt{1 + p_x^2 + p_y^2}. \qquad (6.3)$$

The equilibrium state is determined by the stationarity condition of the total free energy $G = G_b + G_s$. The change of the bulk free energy G_b by the small variation of the interface height $\delta\zeta(x, y)$ is written as

$$\delta G_b = -\frac{\Delta\mu}{\Omega}\int\delta\zeta(x, y)dxdy \equiv -2\lambda\int\delta\zeta dxdy, \qquad (6.4)$$

where we introduce a parameter $\lambda = \Delta\mu/2\Omega$. On the other hand, the surface free energy varies as

$$\delta G_s = \int dxdy\left[\frac{\partial f}{\partial p_x}\delta p_x + \frac{\partial f}{\partial p_y}\delta p_y\right]$$

$$\int dxdy\left[\frac{\partial f}{\partial p_x}\frac{\partial\delta\zeta}{\partial x} + \frac{\partial f}{\partial p_y}\frac{\partial\delta\zeta}{\partial y}\right] = -\int dxdy\delta\zeta\left[\frac{\partial}{\partial x}\frac{\partial f}{\partial p_x} + \frac{\partial}{\partial y}\frac{\partial f}{\partial p_y}\right], \qquad (6.5)$$

where partial integration is performed to derive the last equality. By imposing the stationary condition

$$\delta G_b + \delta G_s = 0 \qquad (6.6)$$

for arbitrary variation $\delta\zeta(x, y)$, one gets the Euler-Lagrange equation

$$2\lambda + \frac{\partial}{\partial x}\frac{\partial f}{\partial p_x} + \frac{\partial}{\partial y}\frac{\partial f}{\partial p_y} = 0. \qquad (6.7)$$

The solution of this equation is

$$\frac{\partial f}{\partial p_x} = -\lambda x, \qquad \text{and} \qquad \frac{\partial f}{\partial p_y} = -\lambda y. \qquad (6.8)$$

It shows that variables x and p_x are conjugate as well as y and p_y are, and $f(p_x, p_y)$ behaves as a potential function. The Legendre transformation introduces the new potential

$$g = f - \frac{\partial f}{\partial p_x}x - \frac{\partial f}{\partial p_y}y = f + \lambda x p_x + \lambda y p_y, \tag{6.9}$$

which is a function of x and y variables as

$$dg = df + d(\lambda x p_x) + d(\lambda y p_y) = \lambda \left(\frac{\partial \zeta}{\partial x}dx + \frac{\partial \zeta}{\partial y}dy\right) = \lambda d\zeta. \tag{6.10}$$

On integration one gets $g(-\lambda x, -\lambda y) = \lambda \zeta$. The potential g is thus equal to the shape ζ as

$$\zeta = \frac{g}{\lambda} = \frac{f}{\lambda} + x p_x + y p_y. \tag{6.11}$$

By rearrangement, it reduces to

$$\frac{\zeta - x p_x - y p_y}{\sqrt{1 + p_x^2 + p_y^2}} = \frac{\gamma(p_x, p_y)}{\lambda}. \tag{6.12}$$

Since the normal vector to the interface is given by $\mathbf{n} = [1 + p_x^2 + p_y^2]^{-1/2}(-p_x, -p_y, 1)$, the point $\mathbf{r} = (x, y, \zeta)$ on the interface satisfies the relation

$$(\mathbf{r} \cdot \mathbf{n}) = \frac{\gamma}{\lambda} = \frac{2\Omega\gamma(\mathbf{n})}{\Delta\mu}. \tag{6.13}$$

The analogous formula holds for a one-dimensional interface of a two-dimensional

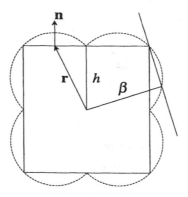

Figure 6.2: Wulff plot and equilibrium shape.

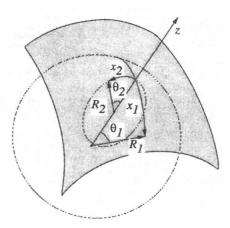

Figure 6.3: Local coordinate system; z axis in normal direction, and x_1 and x_2 in the local principal directions.

system. Actual realization of a one-dimensional interface is a step on a crystal surface with the step free energy $\beta(\mathbf{n})$. The equilibrium shape is determined by the relation

$$(\mathbf{r} \cdot \mathbf{n}) = \frac{\Omega_2 \beta}{\Delta \mu}, \tag{6.14}$$

where $\Omega_2 = a^2$ is an atomic area.

Since the l.h.s. of Eq.(6.13), $h \equiv (\mathbf{r} \cdot \mathbf{n})$, is the vertical length from the crystal center to the interface with orientation $\mathbf{n}$, Eq.(6.13) shows that the length h is proportional to the surface tension $\gamma(\mathbf{n})$. By using this relation one can draw the equilibrium shape of a crystal as shown in Fig.6.2 (Wulff plot) [200]. Draw a line from a center O to the direction $\mathbf{n}$ with a radial length proportional to the surface tension $\gamma(\mathbf{n})$, and then draw a plane perpendicular to it through the end point: The crystal surface can be a part of this plane. By varying the orientation $\mathbf{n}$, the envelope of the perpendicular planes gives the equilibrium shape.

Another way to represent the equilibrium shape is its relation with the interface stiffness. In three dimensions, we choose the local coordinate (x_1, x_2, z) such that z axis is in the normal direction of the interface, and x_1, x_2 axes in the local principal directions with principal radii of curvatures, R_1 and R_2 (Fig.6.3). Then the profile of a crystal near this point is approximately given as

$$z = -\frac{x_1^2}{2R_1} - \frac{x_2^2}{2R_2} \tag{6.15}$$

with $x_1 = R_1\theta_1$, $x_2 = R_2\theta_2$. Here the angles θ_1 and θ_2 are the deviation of the normal direction as shown in Fig.6.3. Then the slopes are $p_i = \partial z/\partial x_i = -x_i/R_i = -\theta_i$ with

$i = 1$ and 2. Equation (6.7) can be expressed as

$$\frac{\Delta\mu}{\Omega} + \sum_{i=1}^{2} \frac{\partial}{\partial x_i} \frac{\partial f}{\partial p_i} = \frac{\Delta\mu}{\Omega} + \sum_{i=1}^{2} \sum_{j=1}^{2} \frac{\partial p_j}{\partial x_i} \frac{\partial^2 f}{\partial p_i \partial p_j} = 0. \tag{6.16}$$

Here $f = \gamma(\theta_1, \theta_2)\sqrt{1 + \theta_1^2 + \theta_2^2}$, and the surface stiffness at $x_1 = x_2 = \theta_1 = \theta_2 = 0$ is

$$\frac{\partial^2 f}{\partial p_i \partial p_j} = \frac{\partial^2 f}{\partial \theta_i \partial \theta_j} \equiv \tilde{\gamma}_{ij} = \gamma \delta_{ij} + \frac{\partial^2 \gamma}{\partial \theta_i \partial \theta_j}. \tag{6.17}$$

The curvature is

$$\frac{\partial p_j}{\partial x_i} = -\frac{1}{R_i} \delta_{ij}. \tag{6.18}$$

Thus Eq.(6.16) reduces to the relation

$$\frac{\Delta\mu}{\Omega} = \frac{\tilde{\gamma}_{11}}{R_1} + \frac{\tilde{\gamma}_{22}}{R_2}. \tag{6.19}$$

If the surface tension is isotropic, then $\tilde{\gamma}_{ij} = \gamma \delta_{ij}$ and the crystal nucleus is spherical with the critical radius

$$R_c = \frac{2\gamma\Omega}{\Delta\mu}, \tag{6.20}$$

obtained from Eq.(6.19) with $R_1 = R_2 = R_c$. This radius corresponds to the critical radius of a three-dimensional nucleus in equilibrium with the undercooling or supersaturation, $\Delta\mu$.

Exercise 1: In two dimensions the interface is defined by the arc length s and the angle θ of the normal vector **n** from y axis. (Fig.6.4)

(1) From Eq.(6.14), show that the point (x, y) on the interface satisfies the relation

$$\lambda x(s) = \beta(\theta) \sin \theta + \beta'(\theta) \cos \theta \tag{6.21}$$

$$\lambda y(s) = \beta(\theta) \cos \theta - \beta'(\theta) \sin \theta \tag{6.22}$$

with $\beta'(\theta) = d\beta/d\theta$ and $\lambda = \Delta\mu/\Omega_2$.

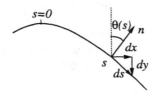

Figure 6.4: Geometry of a step at a site **r** with a normal vector **n**. s is the arc length and θ is the angle of the normal vector.

(2) Find the relation

$$\lambda = \frac{\tilde{\beta}}{\rho} \tag{6.23}$$

with the step stiffness

$$\tilde{\beta} = \beta + \frac{d^2\beta}{d\theta^2} \tag{6.24}$$

and the radius of curvature $\rho = ds/d\theta$.

Answer:

(1) Since $\mathbf{n} = (\sin\theta, \cos\theta)$ from Fig.6.4, Eq.(6.14) can be written as

$$(\mathbf{r} \cdot \mathbf{n}) = x\sin\theta + y\cos\theta = \frac{\beta}{\lambda}. \tag{6.25}$$

By differentiating both sides by s, one obtains

$$\frac{dx}{ds}\sin\theta + \frac{dy}{ds}\cos\theta + (x\cos\theta - y\sin\theta)\frac{d\theta}{ds} = \frac{\beta'}{\lambda}\frac{d\theta}{ds}. \tag{6.26}$$

From the geometry shown in Fig.6.4,

$$dx = ds\cos\theta, \qquad dy = -ds\sin\theta, \tag{6.27}$$

and one gets

$$x\cos\theta - y\sin\theta = \frac{\beta'}{\lambda}. \tag{6.28}$$

Solving (6.25) and (6.28), the desired relations (6.21) and (6.22) are obtained.

(2) Differentiating Eq.(6.28) with s, one gets

$$\frac{dx}{ds}\cos\theta - \frac{dy}{ds}\sin\theta - (x\sin\theta + y\cos\theta)\frac{d\theta}{ds} = \frac{\beta''}{\lambda}\frac{d\theta}{ds}. \tag{6.29}$$

Inserting the values of dx/ds and dy/ds from Eq.(6.27) and by using the relation (6.25), one gets the relation

$$\frac{d\theta}{ds}\frac{\tilde{\beta}}{\lambda} = \cos^2\theta + \sin^2\theta = 1. \tag{6.30}$$

By rearrangement, the desired result (6.23) is obtained.

Exercise 2: Draw the equilibrium shape of a two-dimensional crystal with a tension

$$\beta(\theta) = \beta_0(1 + \tilde{\epsilon}\cos 4\theta). \tag{6.31}$$

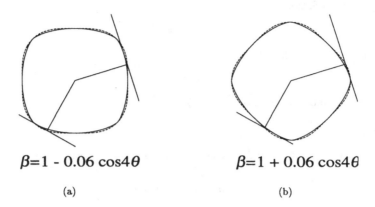

$$\beta=1 - 0.06 \cos 4\theta \qquad\qquad \beta=1 + 0.06 \cos 4\theta$$

(a) \qquad\qquad\qquad\qquad (b)

Figure 6.5: Polar plot of the step free energy $\beta = \beta_0(1 + \tilde{\varepsilon} \cos 4\theta)$ in a dashed curve, and the equilibrium shape in continuous curve. (a) $\tilde{\varepsilon} = 0.06$, and (b) $\tilde{\varepsilon} = -0.06$.

Answer: From Eqs.(6.21-6.22), the shape is determined by the parametric representation;

$$\frac{\lambda}{\beta_0}x = (1 + \tilde{\varepsilon} \cos 4\theta) \sin\theta - 4\tilde{\varepsilon} \sin 4\theta \cos\theta$$

$$\frac{\lambda}{\beta_0}y = (1 + \tilde{\varepsilon} \cos 4\theta) \cos\theta + 4\tilde{\varepsilon} \sin 4\theta \sin\theta. \qquad (6.32)$$

The shape is plotted in Fig.6.5. For $\tilde{\varepsilon} > 0$, the direction $\theta = 0$ corresponds to the maximum of the surface free energy β and to the minimum of the stiffness $\tilde{\beta}$. The equilibrium shape has a pointed corner there because the curvature is large.

Exercise 3: Often a lattice model is used for a simple description of a crystal shape. The space is divided into a regular lattice and each lattice site can be empty or be occupied by a crystal atom. For an occupied site an Ising spin variable $S_i = 1$ is associated, and for an empty site $S_i = -1$. There is cohesion between the neighboring crystal atoms. If the bond from a crystal atom is not connected to the other atom, this broken bond causes an energy cost J. This energetics can be represented by the Ising ferromagnetic Hamiltonian [121]:

$$\mathcal{H} = -\frac{J}{2} \sum_{<ij>} S_i S_j + \frac{NzJ}{4}. \qquad (6.33)$$

Here z is the coordination number, and $z = 4$ for a square lattice model. There the critical point is exactly known to be at [149]

$$\frac{k_B T_c}{J} = \frac{1}{\ln(\sqrt{2}+1)} = 1.1346\cdots \qquad (6.34)$$

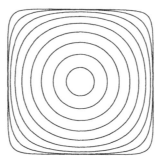

Figure 6.6: Equilibrium shape of 2D Ising model at various temperatures below T_c. Normalized temperatures are $T/T_c=0.1$ (outermost) to 0.9 (innermost) with temperature differences 0.1.

The exact form of the interface free energy $\beta(\theta)$ is also known at a temperature T as a function of the orientation angle θ from the (01) direction as [159]

$$\frac{\beta(\theta)}{k_B T} = |\cos\theta|\sinh^{-1}(\alpha|\cos\theta|) + |\sin\theta|\sinh^{-1}(\alpha|\sin\theta|) \qquad (6.35)$$

with

$$\alpha = \frac{2}{b}\left(\frac{1 - b^2}{1 + \sqrt{\sin^2 2\theta + b^2 \cos^2 2\theta}}\right)^{1/2} \qquad (6.36)$$

and

$$b = \frac{2\sinh(J/k_B T)}{\cosh^2(J/k_B T)}. \qquad (6.37)$$

Draw the equilibrium shape of a square Ising crystal at various temperatures by using this interface energy [159, 7].

Answer: Using the parametric representation, (6.21) and (6.22), the profile is obtained straightforwardly as shown in Fig.6.6. On increasing the temperature the shape becomes isotropic and the size becomes small.

7   Growth Shape

According to the general formalism of nonequilibrium thermodynamics, velocity of the time evolution of a dissipative system is proportional to the thermodynamic driving force given by the gradient of the free energy [142]. Thus the time evolution of the interface $z = \zeta(x, y, t)$ is written by

$$\frac{\partial \zeta(x, y, t)}{\partial t} = -\frac{K\Omega}{k_B T}\frac{\delta G}{\delta \zeta(x, y, t)}. \qquad (7.1)$$

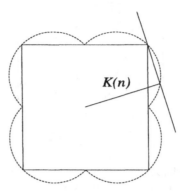

Figure 7.1: Growth shape by Wulff plot using kinetic coefficient K.

In equilibrium Eq.(7.1) reduces to the stationarity condition (6.6) for the equilibrium shape. By choosing the local coordinate such that the z axis is oriented in the interface normal $\mathbf{n}$, one obtains the evolution from Eqs.(6.4–6.5) and (6.16–6.18) as

$$V_n = \frac{K(\mathbf{n})}{k_B T}[\Delta\mu - \Omega(\frac{\tilde{\gamma}_{11}}{R_1} + \frac{\tilde{\gamma}_{22}}{R_2})], \tag{7.2}$$

where we show explicitly the possibility that the kinetic coefficient K depends on the orientation $\mathbf{n}$. The growth law (7.2) is similar to the ideal growth formulae, (3.5),(4.4) and (5.3), but the chemical driving force $\Delta\mu$ is modified by the curvature effect. At the pointed part with positive curvatures, $R_1 > 0$ and $R_2 > 0$, the growth rate decreases by the surface tension effect. This is called the Gibbs-Thomson effect.

For a sufficiently large crystal the radius R_1 and R_2 are very large, and the last term related to the interface stiffness may be neglected. Without it and assuming that $\Delta\mu$ is constant all over the interface, one can easily integrate Eq.(7.2) up to the time t as

$$(\mathbf{r} \cdot \mathbf{n}) = \int_0^t V_n dt = \frac{K(\mathbf{n})\Delta\mu}{k_B T}t, \tag{7.3}$$

where $\mathbf{r}$ is a point on the interface oriented in $\mathbf{n}$ direction. Eq.(7.3) is similar to the equilibrium shape Eq.(6.13), with the interface tension $\gamma(\mathbf{n})$ being replaced by the kinetic coefficient $K(\mathbf{n})$ [46]. Therefore, we can draw the growth shape of a crystal by the Wulff plot of $K(\mathbf{n})$ instead of the $\gamma(\mathbf{n})$, as shown in Fig.7.1. Also one can derive the relation between the local curvature and the kinetic stiffness $\tilde{K}_{ij} = K\delta_{ij} + \partial^2 K/\partial\theta_i\partial\theta_j$. In an isotropic case, the crystal grows spherically with its radius given by

$$R = K\frac{\Delta\mu}{k_B T}t. \tag{7.4}$$

If the orientation **n**−dependence of the kinetic coefficient K is so strong such that a corner appears in the kinetic Wulff plot, the Gibbs-Thomson term with surface tension is no more negligible. The growth shape with both surface tension and kinetics are generally treated by Müller-Krumbhaar et al. [142] and by Uwaha [184].

Part II
Statistical Mechanics of Surface

So far the ideal growth formulae of the crystal are discussed in a macroscopic, thermodynamic sense. In this and the next parts, we consider the microscopic aspects of crystal growth. Since the growth takes place at the crystal surface, it is quite natural to imagine that the surface structure influences the growth. From a microscopic model of the surface configuration, a surface phase transition, a roughening transition [43, 194, 195], is found: At low temperatures surface is flat and smooth, whereas at high temperatures it is rough in the atomic scale. The roughening phase transition induces the faceting transition in the equilibrium crystal shape.

8 Position of Crystallization: Kink Site

Before considering the crystallization in an atomic level, we have to identify when or in what situations an atom is said to be crystallized. In the gas phase an atom is free and makes no bond with other atoms. In a crystal it makes z bonds with neighboring atoms to lower the energy. An energy gain for each connected bond is set $-2J$. For N crystallized atoms, there are totally $zN/2$ bonds, and the total cohesive energy is $-JzN$, or $-Jz$ per each crystal atom. After melting all the atoms become free without any bond connections, and the energy is zero. By assuming that the work associated to the volume change is negligible, the enthalpy variation by the melting or the latent heat per atom is given by $l = zJ$.

If an atom freely moving in the gas phase impinges on the crystal substrate and makes a bond of $z/2$ nearest neighbors with the crystal substrate, there is an energy gain of zJ, and the atom can be regarded to be crystallized. The problem then is when an atom will acquire $z/2$ bonds. On a completely flat crystal surface, the adsorbed atom cannot make so many bonds. But on a crystal surface, there are various defects as shown in Fig.8.1. A flat portion is called a terrace. When the heights of two consecutive terraces differ, there is a step between them. A step can be straight as it runs in a closed packed direction. It can also bent and change the orientation at a kink position. When an atom impinges on a flat terrace, the atom makes only a few bonds with the underlying atoms. The bond connection is so weak that the atom can meander on the surface. This atom is called the adsorbed atom, or adatom for short, and performs surface diffusion. During meandering, the adatom may come in contact with an uprising step. Then step provides some additional bonds parallel to the terrace, but the number of bonds is still in short of $z/2$. The adatom can slide along the step edge by edge diffusion, and may reach to the kink position where there are $z/2$ nearest neighbors. At this kink site, the atom is really said to be crystallized. The striking feature of the kink site is that it never disappears by crystallization or melting; it only slides along the step. For the fast crystallization, therefore, it is

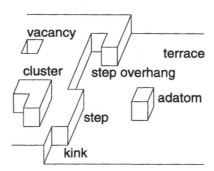

Figure 8.1: Surface configuration with terraces bounded by a step with kinks.

desirable to have many kink sites or steps on the surface. But the surface with many steps and kinks is rough with many broken bonds, and it has high surface energy. The surface cannot be rough at low temperatures where only the minimum energy configuration is allowed. The rough surface is possible only at high temperatures. The roughness of the surface is expected to change drastically at a certain temperature. Since the surface roughness controls the growth mode of the crystal, the roughening phase transition of the surface will be discussed in the next section.

Exercise: In a lattice gas model for crystal growth, a density variable n_i at a lattice site i takes a value 1 when the site is occupied by an atom, or it is 0 when the site is empty. If a broken chemical bond from a crystal atom costs an energy J, the interaction energy is expressed as

$$\mathcal{H} = J \sum_{<ij>} [n_i(1 - n_j) + (1 - n_i)n_j] - \Delta\mu \sum_i n_i, \qquad (8.1)$$

where $<ij>$ means the summation over the nearest neighbor sites pairs, and $\Delta\mu$ is the chemical potential gain by crystallization. Show that the Hamiltonian (8.1) can be transformed to the two-dimensional Ising Hamiltonian

$$\mathcal{H}_\mathrm{I} = -\frac{J}{2} \sum_{<ij>} S_i S_j - H \sum_i S_i. \qquad (8.2)$$

Here an Ising spin variable $S_i = 2n_i - 1$ takes values $+1$ or -1, and the field is $H = \Delta\mu/2$ [128].

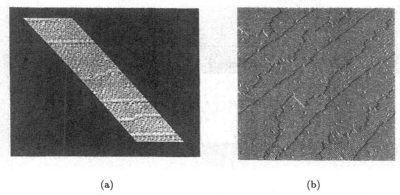

(a)   (b)

Figure 9.1: Surface configuration with steps observed by scanning tunneling microscope (STM). (a) Si(111) [192] and (b) Si(001) faces [178]. (Courtesy of M.G. Lagally).

9   Surface Roughening

We study the equilibrium configuration of the crystal surface in contact with the ambient gas phase, which is realized on the vapor-crystal coexistence curve in equilibrium phase diagram (Fig.2.2). Along this curve the chemical potentials are the same for crystal and gas phases. The temperature variation affects only the surface free energy associated to the surface configuration or its roughness. At low temperatures the crystal is polyhedral, as shown in Fig.1.1, enclosed by atomically flat faces with low Miller indices, called facets, since these faces have low energies. But the flat face has only a single possible configuration, and has no entropy. If atoms jump out from the flat face and sticks on it, there is an entropy gain S connected to the possible numbers of positions for atoms to be taken out and to be put on. But there is an energy cost E, and the equilibrium configuration at a finite temperature T is the one which minimizes the free energy $F = E - TS$. Thus the surface structure is determined by compromising two contributions, E and S [194, 195, 146].

9.1   Monte Carlo Simulation

In order to understand the atomic structure of the crystal surface, it is good to see the surface configuration: *Seeing is believing*. Recent development of various microscopes as scanning tunneling microscope (STM) and others make it possible to see the surface structure in an atomic level, as shown in Fig.9.1. Theoretically, it is also possible to produce atomic configurations of the surface by using various computer simulation methods. One method is the Monte Carlo simulation [25, 76], which produces various microscopic configurations stochastically under the influence of thermal fluctuation. Equilibrium configurations of a surface are obtained by this method as shown in Fig.9.2 [194].

By way of solid-on-solid (SOS) model for the crystal growth (Appendix A9.1),

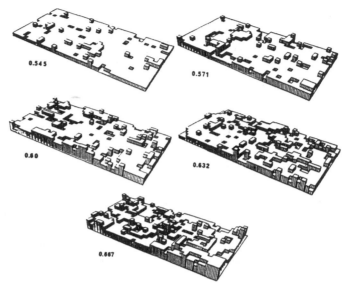

Figure 9.2: Surface configurations by Monte Carlo simulation; Parameters are the temperatures $k_B T / 2J$. [194]

Monte Carlo simulation procedure is now explained [69]. Crystal surface configuration is assumed to be described by the integer height variables $\{h\} \equiv \{h(1), \cdots, h(N)\}$ on a square lattice with $N = L^2$ lattice sites. With the thermal fluctuation, the configuration $\{h\}$ is realized stochastically with a probability $P(\{h\}, t)$ at time t.

The height $h(i)$ at site i changes to another height $h'(i)$ with a transition probability $w(h(i) \to h'(i))$ per time. Then the time variation of the probability P in a small time increment Δt is given by

$$
\begin{aligned}
P(h(1) \cdots h(i) \cdots h(N), t + \Delta t) &= P(h(1) \cdots h(i) \cdots h(N), t) \\
&- \sum_{i=1}^{N} w(h(i) \to h'(i)) \Delta t P(h(1) \cdots h(i) \cdots h(N), t) \\
&+ \sum_{i=1}^{N} w(h'(i) \to h(i)) \Delta t P(h(1) \cdots h'(i) \cdots h(N), t).
\end{aligned} \tag{9.1}
$$

The second term represents the probability decrease due to the probability escape from $\{h\}$ to other configurations, and the third term represents the increase due to the probability income from other configurations to $\{h\}$. In the continuous time limit,

$\Delta t \to 0$, one obtains the master equation of the probability;

$$\frac{\partial P\left(\{h\},t\right)}{\partial t} = -\sum_{i=1}^{N} w\left(h(i) \to h'(i)\right) P\left(\{h\},t\right) + \sum_{i=1}^{N} w\left(h'(i) \to h(i)\right) P\left(\{h\}_i,t\right).$$

(9.2)

Here $\{h\}_i$ is a configuration different from $\{h\}$ only by a height at the site i. In order to realize thermal equilibrium asymptotically ($t \to \infty$), the Boltzmann distribution

$$P_{\text{eq}} = Z^{-1} \exp\left[-\frac{\mathcal{H}(\{h\})}{k_B T}\right]$$

(9.3)

should be a stationary solution of the master equation. Here $\mathcal{H}(\{h\})$ is the Hamiltonian defined as

$$\mathcal{H}(\{h\}) = J \sum_{<ij>} |h(i) - h(j)| + \Delta\mu \sum_i h(i) = E(\{h\}) + \Delta\mu \sum_i h(i),$$

(9.4)

with J being the energy cost of a broken bond, $\Delta\mu = \mu_{\text{G}} - \mu_{\text{S}}$ the chemical potential difference between the gas and the crystal phases. If the transition probability w satisfies the detailed balance condition;

$$\frac{w(h(i) \to h'(i))}{w(h'(i) \to h(i))} = \frac{P_{\text{eq}}(\{h\}_i)}{P_{\text{eq}}(\{h\})} = \exp\left[-\frac{\mathcal{H}(\{h\}_i) - \mathcal{H}(\{h\})}{k_B T}\right],$$

(9.5)

the equilibrium is confirmed. The crystallization rate at the site i is chosen to depend only on the chemical potential difference as

$$w(h(i) \to h(i) + 1) = \exp\left(\frac{\Delta\mu}{k_B T}\right).$$

(9.6)

The evaporation rate at the site i is determined from the detailed balance condition (9.5) as

$$w(h(i) \to h(i) - 1) = \frac{P_{\text{eq}}(h(i) - 1)}{P_{\text{eq}}(h(i))} w(h(i) - 1 \to h(i)) = \exp\left(-\frac{\Delta E}{k_B T}\right).$$

(9.7)

Here $\Delta E = E(h(i) - 1) - E(h(i))$ is the variation of the bond energy associated to the lowering of the height $h(i)$, and is given by $\Delta E = -2J(n - z/2)$ with n being the number of nearest neighbor sites with heights lower than $h(i)$ and z is the coordination number. Surface configurations with corresponding n are shown in Fig.9.3 for a one-dimensional example ($z = 2$). Since the maximum of n is z, the maximum evaporation rate is proportional to $e^{zJ/k_B T}$. Thus by choosing

$$\Delta t = \frac{1}{e^{\Delta\mu/k_B T} + e^{zJ/k_B T}},$$

(9.8)

the transition rate $w\Delta t$ in Eq.(9.1) never exceeds unity.

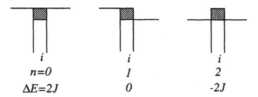

Figure 9.3: Configurations of a one-dimensional interface ($z = 2$) before evaporation of the shaded top atom on the i-th column. n means the number of nearest neighbor sites with the height lower than h_i, and ΔE is the variation of the bond energy.

The real configuration change of the simulation is performed as follows: Select a lattice site i randomly, and calculate n and various transition probabilities $w\Delta t$. Then, create a random number u which is uniformly distributed between 0 and 1. If u lies between 0 and $\Delta t e^{\Delta\mu/k_B T}$, the height $h(i)$ is increased by 1. If u lies between $\Delta t e^{\Delta\mu/k_B T}$ and $\Delta t(e^{\Delta\mu/k_B T} + e^{-\Delta E/k_B T})$, the height $h(i)$ is decreased by 1. Otherwise, the height does not change. After trying the height variation N times, every site has been tried once on average to change its height, and the one Monte Carlo step (MCS) is said to have been performed. The time increases by Δt. See the sample program in Appendix A9.2.

The average height at the m-th MCS is defined as

$$\langle h(m) \rangle = \frac{1}{N} \sum_{i=1}^{N} h(i; m), \tag{9.9}$$

where $h(i; m)$ represents the height at a site i at m-th MCS. In equilibrium where $\Delta\mu = 0$, the height should remain constant, and the thermal average of height is obtained by

$$\langle h \rangle = \frac{1}{M} \sum_{m=M_0}^{M_0+M} \langle h(m) \rangle. \tag{9.10}$$

Other quantities are calculated similarly.

In some cases, the Monte Carlo procedure can be regarded not merely a mathematical algorithm to obtain equilibrium quantities, but also as a physical process of relaxation dynamics. For example, the SOS simulation so far explained can be regarded as to mimic a crystal growth from a homogeneous ambient gas phase. In order to incorporate the kinetic coefficient K, the time increases $\Delta t/K$ after one MCS. The growth rate (5.2) with additional curvature correction is obtained from the simulation by

$$V(t) = \frac{K}{\Delta t} \left(\langle h(m+1) \rangle - \langle h(m) \rangle \right), \tag{9.11}$$

where the time is $t = m\Delta t/K$.

At very low temperatures ($k_B T \ll J$) with a small supersaturation, the time increment Δt defined in Eq.(9.8) is so small that configuration change occurs very seldom. In order to circumvent the difficulty and alter the configuration efficiently, the waiting time method is often used. Details are found in references [69, 188, 88].

Simulation results of the equilibrium configuration are shown in Fig.9.2 [70]. At low temperatures, the surface is flat with few bumps and holes, whereas at high temperatures the surface is rough. Thus the change of the surface configuration by temperature is obvious. This surface structure change is in fact a phase transition, which takes place at a critical temperature. For the real understanding of the phase transition, one needs analytical theory on it.

9.2   Mean Field Theory by Jackson; α-Parameter

For an analytical study on the roughening transition, we start from the simplest model, where the interface is assumed to consist of a single layer between the semi-infinite crystal and the vacuum phases [43, 90]. The interface layer is divided into a two dimensional lattice cells, and each cell site can be occupied by a crystal atom or can be empty. Between the neighboring crystal atoms a cohesive energy of attraction is assumed.

If the layer is almost empty, the layer belongs to the vacuum phase and the crystal phase terminates sharply before the interface layer: The crystal has a sharp interface. If the interface layer is almost fully occupied, the layer belongs to the crystal, and the crystal terminates sharply after the interface layer. Only when the half of the layer is empty and the other half is occupied by crystal atoms, the interface layer belongs neither to the crystal nor to the vacuum phase. In this case the interface can be regarded to be rough. Burton, Cabrera and Frank treated the square lattice case exactly [43], because the model is equivalent to the exactly soluble Ising ferromagnet as (8.1) or (8.2). Jackson later treated the same model in the mean field approximation [90], and estimated the roughening temperature T_R, and compared it with the melting temperature T_M. Here I summarize briefly the latter mean-field treatment.

The interface layer consists on N cell sites, and each cell site has z_s nearest neighbor sites in the layer. When there are N_S crystal atoms on the interface layer, the energy increase is estimated as follows: Among z_s neighbors of each crystal atom, $(1 - N_S/N)z_s$ of them are expected to be empty in the mean field approximation. Therefore, the total energy cost E is proportional to the total number of broken crystal bonds as

$$E = J z_s (1 - N_S/N) N_S. \qquad (9.12)$$

The entropy S is obtained as the logarithm of the number W of the microscopically different configurations. Since W is the number of ways to choose N_S sites from N lattice points to locate crystal atoms, it is easily obtained as $W = N!/[N_S!(N - N_S)!]$. By using the Stirling's formula $\ln N! \approx N \ln N - N$, the entropy is obtained as

$$S = k_B \ln W \approx k_B [N \ln N - N_S \ln N_S - (N - N_S) \ln(N - N_S)]. \qquad (9.13)$$

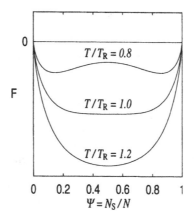

Figure 9.4: Free energy F as a function of the parameter $\Psi = N_S/N$ at various temperatures. Below the roughening temperature T_R, F has double minima, whereas above T_R it has a single minimum.

The dependence of the free energy $F = E - TS$ on the parameter $\Psi = N_S/N$ is shown in Fig.9.4 at various temperatures. By minimizing the free energy F, we obtain the equilibrium value of the parameter Ψ, or the degree of crystallinity of the interface. At a temperature below the roughening temperature

$$T_R = \frac{z_s J}{2k_B},\qquad(9.14)$$

there are two minimum points in $F - \Psi$ diagram, showing that the interface is almost empty ($\Psi \ll 1$) or is almost full ($\Psi \approx 1$). At high temperatures there is a single minimum at the intermediate value $\Psi = 0.5$; The interface is rough.

Since the coupling energy J is related to the latent heat per molecule by $l = zJ$ with z being the bulk coordination number, the roughening temperature can be expressed as

$$T_R = \frac{z_s l}{2z k_B}.\qquad(9.15)$$

The comparison of the melting temperature T_M and the roughening one T_R yields the following parameter

$$\alpha \equiv \frac{z_s l}{z k_B T_M} \left(= \frac{2T_R}{T_M} \right),\qquad(9.16)$$

which is readily evaluated by using experimental data. If $\alpha < 2$ the melting point T_M is higher than the roughening temperature T_R and the interface at T_M should be rough. If $\alpha > 2$, on the other hand, T_M is below the roughening temperature and the interface is flat. The parameter α is called the Jackson's α parameter [90], and is used to classify the material for surface roughening or to guide experiments.

This model is equivalent to two dimensional Ising ferromagnet defined in (6.31), and the exact solution is known [149]. The singularities in thermodynamic quantities are well studied, and for example, the specific heat shows a logarithmic divergence. But the single layer model of interface cannot correctly describe the roughening transition. In this model the height difference at two separated points is 0 or 1 by definition, and the interface is always flat. If the interface is really rough, heights at two distant points should be uncorrelated and fluctuate strongly to diverge at infinite separation. Burton, Cabrera and Frank extended the model with multiple layers and analyzed it in the Bethe approximation [43]. They found no singularity in the free energy, meaning the absence of phase transition when the height fluctuation increases. The Bethe approximation is better than a simple mean field approximation, but as for critical phenomena the same qualitative behavior is expected since both approximations neglect the effect of fluctuation. For the analysis of the true roughening transition, one should include the possibility of arbitrary height difference at two separated points. The thermal fluctuation influences strongly on the phase transition, and one needs sophisticated method as variational [163] or renormalization group method [114, 49, 146]. The critical behavior of the phase transition turns out to be quite different from that of the single-layer Ising model. However, the transition temperature is quantitatively not much different from the two-dimensional Ising model, and thus the Jackson's α-parameter is used as a convenient criterion for surface roughening.

9.3   Mean Field Argument in terms of Steps

We now look the surface roughening from another point of view, in terms of steps on a crystal surface. For a rough surface there are two- dimensional nuclei, where the inside terrace is one atomic height higher or lower than the exterior one, as shown in Fig.9.5. For a nucleus with the perimeter step length L, the energy cost is $E = JL/a$ due to the broken bonds. Here a is the linear size of the atom.

As for the entropy, one counts the possibility of step configurations with a fixed perimeter length L. A step looks quite similar to a polymer with L/a segments. Each segment should be oriented in one of the z_s directions of nearest neighbor connection. The first segment can be put in one of z_s possible orientations. From the second segment on, it can be in one of $z_s - 1$ orientations, because it cannot fold back on the previous segment. Then the entropy is approximately given by $S \approx k_B \ln(z_s - 1)^{L/a} = k_B(L/a)\ln(z_s - 1)$. Here we have neglected such restrictions that the step loop should be closed and cannot cross itself. The free energy is then obtained as $F = [J - k_B T \ln(z_s - 1)](L/a) \equiv \beta L$, with β being a step free energy per length. At a low temperature below the roughening temperature $T_R = J/k_B \ln(z_s - 1)$, β is positive and the state which minimizes the free energy F contains no step ($L = 0$): The surface is flat. At a high temperature ($T > T_R$) $L = \infty$ corresponds to the state of the minimum free energy and the surface becomes rough with an infinitely long step running. In this case the step-step interaction becomes important and the step

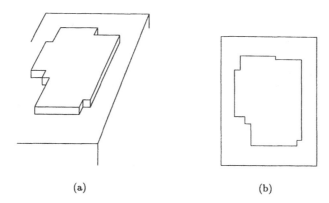

(a)   (b)

Figure 9.5: (a) A two-dimensional nucleus bounded by a step with length L. (b) Top view of a step loop.

free energy is shown to remain zero.

9.4 Short Summaries of Roughening Transition

In order to understand the roughening phase transition correctly, the mean field approximation is insufficient and the fluctuation effect has to be considered carefully. Treatments by the variational method [163] and by the renormalization group theory [114, 49, 145, 146] are described in Appendices A9.3-4 in detail, and here we give a short summary of the renormalization results. There are also some exactly soluble models of the surface structure [187, 93] which shows the roughening transition, as is described in Appendix A9.5. The result is also summarized here.

In the SOS model the interfacial configuration is described by the height variable $h(\mathbf{r})$ at a two-dimensional site $\mathbf{r}$. Below the roughening temperature T_R the interface is flat and the height fluctuation is small such that the height difference correlation

$$G(\mathbf{r}) \equiv \langle (h(\mathbf{r} + \mathbf{r}_0) - h(\mathbf{r}_0))^2 \rangle \tag{9.17}$$

at two separate points remains finite. On the other hand, above T_R the interface is rough and the height fluctuation is so large that $G(r)$ diverges as $r \to \infty$. According to the renormalization group calculation [146], $G(r)$ diverges logarithmically above T_R. Below T_R the height fluctuate logarithmically at short distances within the correlation length ξ, but for large distances $r > \xi$ the fluctuation remains constant:

$$G(r) \sim \begin{cases} \ln r, & \text{for } r < \xi \\ \ln \xi, & \text{for } r > \xi \,. \end{cases} \tag{9.18}$$

The correlation length diverges at the roughening point T_R as

$$\xi \sim \exp[\frac{C}{\sqrt{T_R - T}}]. \tag{9.19}$$

Within the scale of correlation length ξ, the surface is rough, as Eq.(9.18) shows. This means that steps with a perimeter length up to ξ can be excited by thermal fluctuation. By denoting the step free energy per length as β, the total step free energy $\beta\xi$ of a step with a length ξ should be of the order of the thermal energy $k_B T$. Therefore, the step free energy β is given by

$$\beta \sim \xi^{-1} \sim \exp[-\frac{C}{\sqrt{T_R - T}}]. \tag{9.20}$$

The singularity of the step free energy is an essential one, whose derivatives of arbitrary order vanish at the transition point T_R and there is no divergence in derivatives, such as in specific heat. Above T_R, the correlation length ξ is infinite and thus $\beta = 0$.

There are some special models of a crystal surface, which are exactly soluble [187, 93]. They are mapped onto the inverted F-model, a special case of a six-vertex model [129]. The step free energy is shown to have the same singularity with that predicted by the renormalization group method, Eq.(9.20).

The Monte Carlo simulations also show that the height difference correlation function $G(r)$ diverges logarithmically above T_R, and it saturates below T_R [127, 164].

10 Step Fluctuation in Equilibrium

Above the roughening temperature the crystal surface is rough and there are kinks everywhere on the surface. In this case the growth formula of Hertz and Knudsen may be valid, if the transport is sufficiently fast. Below T_R the surface is flat and there is no thermally excited kinks. Then, how can the crystal grow?

When an atom impinges on a flat surface from the ambient gas phase as shown in Fig.8.1, the adatom moves around the surface before it evaporates back into the gas. If an adatom meets other adatoms during surface diffusion, they make bonds and form a cluster to gain energy. The cluster is stabler than the isolated adatom. Further coalescence of adatoms makes the cluster larger to form a nucleus as shown in Fig.9.5. Around this two-dimensional cluster or crystal nucleus there is a step and kink positions.   Thus the two-dimensional nucleus provides the kink sites for the crystal growth. This mechanism of growth is called the two-dimensional nucleation and growth.

There is another mechanism of crystal growth. The molecular arrangement of the crystal is not always perfect, but sometimes there are defects. Among them is the screw dislocation, a line defect where the molecular arrangement is shifted in two consecutive lattice planes. When the screw dislocation ends up at the crystal surface, it produces a step running from the dislocation line (Fig.10.1). If there are many

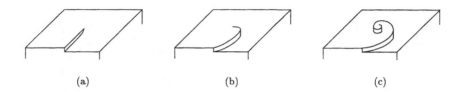

(a) (b) (c)

Figure 10.1: Spiral growth by a screw dislocation.

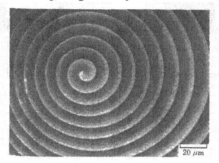

Figure 10.2: Spiral on SiC.(Courtesy of I.Sunagawa.)

kinks on a step, the meandering adatoms meet the step, are incorporated in the kink sites, and the step advances. But since one end of the step is pinned by the screw dislocation, the step winds up around the dislocation in a spiral form. This spiral step never disappears but grows to supply the kink position. This mode of growth is called the spiral growth (Fig.10.1 and 10.2). The nucleation and spiral growths are two main mechanisms for the growth of the crystal with flat faces. Growth laws of the two modes, nucleation and spiral growths, will be described later.

First we calculate the kink density on the step. Since the kinks are related to the change of orientation of the step, it is related to the step fluctuation, too. Nowadays step fluctuation can be observed by various atomic-level microscopes as scanning tunneling microscope (STM) [178, 192, 86, 41], atomic force microscope (AFM), reflection electron microscope (REM) [4, 125, 126] and so on. STM and REM observations are shown in Fig.9.1 and Fig.10.3, respectively. From these observations one can derive the microscopic information of step stiffness and energy parameters.

Let us consider a simple model of the step running, on average, in x direction, as shown in Fig.10.4. By denoting a lattice constant a, the step position is written as $y(i)$ at $x = ia$. When the step runs in the next position $x = (i + 1)a$, its height $y(i+1)$ can be the same with the previous one, or can be up or down by a single unit and form a kink. The total energy of a step running from $x = 0$ to $x = L$ is written

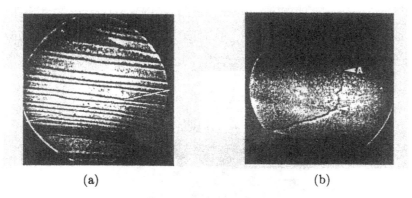

(a)   (b)

Figure 10.3: (a) A step train and (b) an isolated step observed by reflection electron microscope (REM) [4].

as

$$\mathcal{H} = J_y \frac{L}{a} + J_x \sum_{i=1}^{L/a} \frac{|y(i) - y(i-1)|}{a}, \qquad (10.1)$$

with J_x, J_y being the energy cost of a broken bond in x and y directions respectively. Since the height difference $(y(i) - y(i-1))/a$ between neighboring positions takes only three values, 0 or ± 1, one can easily calculate the partition function at a temperature T as

$$Z = \left[\sum_{n=-1}^{+1} \exp\left(-\frac{J_y + J_x|n|}{k_B T}\right) \right]^{L/a} = \exp\left(-\frac{L J_y}{a k_B T}\right) \left[1 + 2\exp(-\frac{J_x}{k_B T}) \right]^{L/a} . \qquad (10.2)$$

Therefore, the step free energy per length is obtained as

$$\beta_0 = -\frac{k_B T \ln Z}{L} = \frac{1}{a}\left(J_y - k_B T \ln\left[1 + 2\exp\left(-\frac{J_x}{k_B T}\right) \right] \right). \qquad (10.3)$$

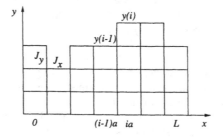

Figure 10.4: Solid-on-solid model of a step running on average in x-direction.

The density of a kink is also obtained as

$$\langle |n| \rangle = \frac{2 \exp(-J_x/k_B T)}{1 + 2 \exp(-J_x/k_B T)}. \tag{10.4}$$

J_x thus corresponds to the kink energy. At a finite temperature, there is a finite density $\langle |n| \rangle$ of kinks on a step, and thus the adsorbed atoms reached to the step can be easily crystallized at kink sites. If $J_x = J_y = J$, the step free energy density β_0 vanishes at a temperature $T_R = J/(k_B \ln 2)$. This is said to be the roughening transition temperature of this model.

The correlation function of the step height difference at a separation x defined as

$$G(x) \equiv \langle (y(x) - y(0))^2 \rangle = \langle (\sum_{i=1}^{x/a} [y(i) - y(i-1)])^2 \rangle = \langle (\sum_{i=1}^{x/a} a n_i)^2 \rangle \tag{10.5}$$

characterizes the roughness of the step. Since the height difference $a n_i$ at two neighboring sites i and $i-1$ has no correlation with those at other positions, the step fluctuation is easily calculated as

$$G(x) = a^2 \sum_{i=1}^{x/a} \langle n_i^2 \rangle = a x \langle n^2 \rangle. \tag{10.6}$$

The height correlation diverges in proportional to the separation x as x increases, and the step is rough at any finite temperatures. In the present model, $\langle n^2 \rangle = \langle |n| \rangle$ and the coefficient in Eq.(10.6) is proportional to the kink density (10.4).

We can relate the step stiffness $\tilde{\beta}$ of Eq.(6.24) to the step fluctuation. In a coarse-grained continuum model, the total step free energy is written as

$$F_{\text{step}} = \int ds \beta \left(\theta(s) \right) = \int dx \beta \left(\theta(s) \right) \sqrt{1 + \left(\frac{dy}{dx} \right)^2}. \tag{10.7}$$

Here s denotes the arc length along the step, and $\theta(s)$ is the angle between the normal vector $\mathbf{n}$ and y axis as $\tan \theta = -dy/dx$. See Fig.6.4. As the step is rough, the step free energy β is expected to have no singularity in the orientation θ-dependence. Thus, for a small step fluctuation, one can approximate $\theta- \approx dy/dx$, $\sqrt{1 + (dy/dx)^2} \approx 1 + \frac{1}{2}(dy/dx)^2$, $\beta(\theta) = \beta_0 + \frac{1}{2}(d^2\beta/d\theta^2)\theta^2$. Then the integrand in Eq.(10.7) can be expanded up to the second order of the derivative dy/dx as

$$F_{\text{step}} \approx \int dx [\beta_0 + \frac{1}{2}\tilde{\beta}(\frac{dy}{dx})^2], \tag{10.8}$$

where $\tilde{\beta} \equiv \beta + \beta''$ is the step stiffness in $\theta = 0$ direction [60, 61]. From this deformation free energy F_{step}, the thermal expectation of the slope fluctuation is easily calculated by the equipartition as $\langle (dy/dx)^2 \rangle = k_B T/\tilde{\beta}$. Since the slope at different positions

are uncorrelated, the height difference correlation function $G(x)$ is easily calculated
as

$$G(x) = \left\langle \left(\int_0^x \frac{dy}{dx} dx \right)^2 \right\rangle = \int_0^x \left\langle \left(\frac{dy}{dx} \right)^2 \right\rangle dx = \frac{k_B T}{\tilde{\beta}} x. \tag{10.9}$$

Comparison of Eq.10.9) with the previous microscopic result (10.6) indicates that the
stiffness $\tilde{\beta}$ is related to the height difference fluctuation as

$$\tilde{\beta} = \frac{k_B T}{a \langle n^2 \rangle} = \frac{k_B T}{2a} \frac{1 + 2 \exp(-J_x/k_B T)}{\exp(-J_x/k_B T)}. \tag{10.10}$$

We choose the zero of the y coordinate to the average position of the step as
$\langle y(x) \rangle = 0$. Then the step fluctuation is characterized by the width w defined by
$w^2 = \langle y(x)^2 \rangle$. In order to calculate w, it is convenient to transform the step fluctuation
in Fourier modes:

$$y(x) = \sum_q y(q) e^{iqx}, \tag{10.11}$$

where $q = 2\pi m/L$ with the integer $m = -\infty, \cdots, \infty$. Here we assume the periodic
boundary condition: $y(0) = y(L)$. The mode amplitude is obtained by

$$y(q) = \frac{1}{L} \int_0^L y(x) e^{-iqx} dx = y^*(-q). \tag{10.12}$$

The step free energy F_{step} is written in terms of the Fourier modes as

$$F_{\text{step}} = \beta_0 L + \frac{L\tilde{\beta}}{2} \sum_q q^2 |y(q)|^2. \tag{10.13}$$

The thermal average of the fluctuation of each mode is obtained from equipartition
as

$$\langle |y(q)|^2 \rangle = \frac{k_B T}{L\tilde{\beta} q^2}, \tag{10.14}$$

and the step fluctuation width w is calculated as

$$w^2 = \sum_q \langle |y(q)|^2 \rangle = \frac{k_B T}{L\tilde{\beta}} \sum_q \frac{1}{q^2} = \frac{k_B T}{L\tilde{\beta}} \left(\frac{L}{2\pi} \right)^2 2 \sum_{m=1}^{\infty} \frac{1}{m^2} = \frac{k_B T}{12\tilde{\beta}} L. \tag{10.15}$$

Reflecting the roughness of the step, its width w diverges for a long step as $L^{1/2}$.

Exercise 1: Show that a step pinned at both ends, $y(0) = y(L) = 0$, fluctuates twice
larger than a periodic step, and has the step fluctuation

$$w^2 = \frac{k_B T}{6\tilde{\beta}} L. \tag{10.16}$$

The formula can be used to analyze the fluctuation of an isolated step shown in Fig.10.3b to obtain the step stiffness $\tilde{\beta}$ [4].

Answer: Step fluctuation is decomposed into sine-modes as

$$y(x) = \sum_q y(q) \sin qx \tag{10.17}$$

with $q = \pi m/L$ and m is the positive integer as $m = 1, 2, \cdots, \infty$. The amplitude $y(q)$ is obtained by

$$y(q) = \frac{2}{L} \int_0^L y(x) \sin qx \, dx. \tag{10.18}$$

From the ortho-normal condition

$$\frac{2}{L} \int_0^L \sin qx \sin q'x \, dx = \frac{2}{L} \int_0^L \cos qx \cos q'x \, dx = \delta_{qq'}, \tag{10.19}$$

the step free energy is written as

$$F_{\text{step}} = \frac{\tilde{\beta}}{4} L \sum_q q^2 y(q)^2. \tag{10.20}$$

From the equipartition, the thermal average of the amplitude fluctuation is obtained as

$$\langle y(q)^2 \rangle = \frac{2k_B T}{L \tilde{\beta} q^2}. \tag{10.21}$$

The step width w is then calculated as

$$\begin{aligned}
w^2 &= \frac{1}{L} \int_0^L \langle y(x)^2 \rangle dx = \frac{1}{2} \sum_q \langle y(q)^2 \rangle \\
&= \frac{k_B T}{L \tilde{\beta}} \left(\frac{L}{\pi} \right)^2 \sum_{m=1}^{\infty} \frac{1}{m^2} = \frac{k_B T}{6 \tilde{\beta}} L. \tag{10.22}
\end{aligned}$$

Exercise 2: If the tilting field H_t is applied on the step model (10.1), the Hamiltonian is written as

$$\mathcal{H} = J_y \frac{L}{a} + J_x \sum_{i=1}^{L/a} \frac{|y(i) - y(i-1)|}{a} - H_t \frac{y(L/a) - y(0)}{a}, \tag{10.23}$$

and the step is tilted such that the height at the right end $\langle y(L/a) \rangle$ differs from the fixed height at the left end $y(0)$.

(1) Calculate the free energy $\tilde{F}$ as a function of H_t.

(2) Calculate the average slope $p = \langle y(L/a) - y(0) \rangle / L = \langle n \rangle$ as a function of H_t, and show that for small H_t slope p is proportional to H_t as

$$p = \frac{H_t}{\tilde{\beta} a}, \tag{10.24}$$

with the step stiffness obtained by Eq.(10.10). The stiff step resists tilting against the tilting field.

(3) Show that the free energy of a tilted step with a slope p defined by the Legendre transformation $F(p) = \tilde{F}(H_t) + LH_t p/a$ behaves as

$$F(p) = F(0) + \frac{L}{2}\tilde{\beta}p^2 \qquad (10.25)$$

for small slope p.

Answer:

(1) By defining $n_i = (y(i) - y(i-1))/a$, the height difference at both ends is represented as $y(L/a) - y(0) = a\sum_{i=1}^{L/a} n_i$. The partition function is calculated as

$$
\begin{aligned}
Z &= \left[\sum_{n=-1}^{+1} \exp\left(-\frac{J_y + J_x|n| - H_t n}{k_B T}\right)\right]^{L/a} \\
&= \exp\left(-\frac{LJ_y}{ak_B T}\right)\left[1 + 2\exp\left(-\frac{J_x}{k_B T}\right)\cosh\frac{H_t}{k_B T}\right]^{L/a}. \qquad (10.26)
\end{aligned}
$$

The free energy $\tilde{F}$ is then obtained as

$$\tilde{F} = -k_B T \ln Z = \frac{L}{a}\left(J_y - k_B T \ln\left[1 + 2e^{-J_x/k_B T}\cosh\frac{H_t}{k_B T}\right]\right). \qquad (10.27)$$

(2) The average of the height difference is calculated as

$$\langle y(L/a) - y(0)\rangle = -a\frac{\partial \tilde{F}}{\partial H_t} = L\frac{2e^{-J_x/k_B T}\sinh(H_t/k_B T)}{1 + 2e^{-J_x/k_B T}\cosh(H_t/k_B T)}. \qquad (10.28)$$

The slope is then obtained as

$$p = \frac{2e^{-J_x/k_B T}\sinh(H_t/k_B T)}{1 + 2e^{-J_x/k_B T}\cosh(H_t/k_B T)} \approx \frac{H_t}{k_B T}\frac{2e^{-J_x/k_B T}}{1 + 2e^{-J_x/k_B T}} = \frac{H_t}{\tilde{\beta}a}. \qquad (10.29)$$

(3) Since $Lp = -a\partial\tilde{F}(H_t)/\partial H_t$, the free energy $F(p)$ is calculated as

$$
\begin{aligned}
F(p) &= \tilde{F}(H_t) - H_t\frac{\partial\tilde{F}(H_t)}{\partial H_t} \approx \tilde{F}(0) + H_t\tilde{F}'(0) + \frac{1}{2}H_t^2\tilde{F}''(0) \\
&\quad - H_t\left[\tilde{F}'(0) + H_t\tilde{F}''(0)\right] = F(0) + \frac{L}{2}\tilde{\beta}p^2. \qquad (10.30)
\end{aligned}
$$

In the last equality we used the approximation (10.29) for small p and the relation $\tilde{F}''(0) = -(L/a)\partial p/\partial H_t = -L/\tilde{\beta}a^2$. The result (10.30) agrees with the free energy of a step with a constant slope $p = dy/dx$ calculated from Eq.(10.8).

Exercise 3: Consider the similar lattice model for the step but the step height difference between the neighboring sites, $n = (y(i+1) - y(i))/a$, can take arbitrary integer values from $-\infty$ to ∞. Then calculate the step free energy and the roughening temperature. This roughening temperature agrees in fact with an exact transition temperature of a square Ising ferromagnet. Also calculate the height difference correlation function $\langle n^2 \rangle$ and the step stiffness $\tilde{\beta}$.

Answer: The partition function is calculated as

$$Z = \exp\left(-\frac{LJ_y}{ak_BT}\right) \left[\sum_{-\infty}^{\infty} \exp\left(-\frac{|n|J_x}{k_BT}\right)\right]^{L/a} = \exp\left(-\frac{L\alpha_y}{a}\right) \left(\coth\frac{\alpha_x}{2}\right)^{L/a}, \quad (10.31)$$

where the parameters α_x and α_y are defined as

$$\alpha_{x,y} = \frac{J_{x,y}}{k_BT}. \quad (10.32)$$

Then the step free energy per length β_0 is obtained as

$$\frac{\beta_0 a}{k_BT} = -\frac{\ln Z}{L/a} = \alpha_y - \ln\coth\frac{\alpha_x}{2}. \quad (10.33)$$

At the roughening point, $\beta_0 = 0$ and thus

$$\alpha_{y,R} = \ln\coth\frac{\alpha_{x,R}}{2}. \quad (10.34)$$

It is straightforward to transform (10.33) in a symmetric form as

$$\sinh\alpha_{x,R}\sinh\alpha_{y,R} = 1. \quad (10.35)$$

This is the relation to give the transition temperature of the square Ising ferromagnet with the nearest neighbor interactions, J_x and J_y in x and y directions respectively [193]. For an equal coupling, $J_x = J_y = J$, the transition temperature is at $T_R = J/k_B \ln(\sqrt{2}+1)$, in agreement with Eq.(6.34).

Fluctuation of a step is calculated as

$$\langle n^2 \rangle = \frac{2e^{-\alpha_x}}{(1 - e^{-\alpha_x})^2} = \frac{1}{2\sinh^2\alpha_x/2}, \quad (10.36)$$

and the stiffness as

$$\tilde{\beta} = \frac{k_BT}{a\langle n^2 \rangle} = \frac{2k_BT}{a}\sinh^2\frac{\alpha_x}{2}. \quad (10.37)$$

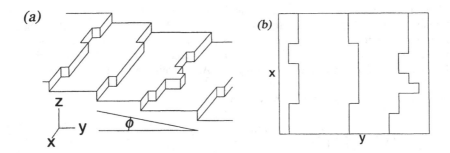

Figure 11.1: (a) Vicinal face and (b) the top view of step configuration.

11   Step-Step Interaction induced by Entropy

Consider now a surface with a small inclination, which is called a vicinal face. A vicinal face consists of terraces of a low indexed singular face and steps separating these terraces, as shown in Fig.11.1.   We consider for simplicity that steps are running in x direction with a mutual separation ℓ. The slope of this vicinal face is then $|\partial z/\partial y| = a/\ell$, where a is the step height. The surface normal makes an angle ϕ with z-axis:

$$|\phi| = \arctan\frac{a}{\ell} \approx \frac{a}{\ell} \tag{11.1}$$

for a small tilting. In a unit projected area in xy-plane, there are $1/\ell$ steps running and the surface free energy f is written as

$$f(\phi) = \frac{\gamma(\phi)}{\cos\phi} = \gamma_0 + \frac{\beta(\phi)}{\ell}, \tag{11.2}$$

where $\gamma(\phi)$ is the surface free energy per unit area, γ_0 is that of a flat terrace, and $\beta(\phi)$ is the step free energy per length for a vicinal face with a tilting ϕ. At the absolute zero temperature, $T = 0$, the step runs straight and βa is equal to J_y of the energy of the broken bond. At a finite temperature, a single step fluctuates and the step free energy decreases as Eq.(10.3). For a vicinal face, there are many other steps and they cannot cross with each other because of the large energy cost for a configuration with overhangs. Thus the fluctuation of a step is limited by the neighboring steps, and the entropy is reduced, or the step free energy is enhanced.

We consider the same step model proposed in the previous section. Since the kink formation costs an additional energy J_x, the probability that a step meanders plus or minus y direction is equal to the Boltzmann weight,

$$p_K = \frac{\exp(-J_x/k_B T)}{1 + 2\exp(-J_x/k_B T)}. \tag{11.3}$$

Then meandering of a step in y-direction can be regarded as a trajectory of a one dimensional Brownian motion with a "time" x/a: At each time step Brownian particle meanders to plus or minus in y directions with a probability p_K and remains there with a probability $1 - 2p_K$. Then after x/a time steps, particle deviates from the original position about $(\Delta y/a)^2 \sim (x/a)p_K$. If the step fluctuates a separation ℓ, then it collides with the other steps on average. The collision takes place at every length of $\Delta x \sim a p_K^{-1}(\ell/a)^2$. This means in a unit length there are $(\Delta x)^{-1}$ collisions, and the step looses the entropy. The step free energy increases by about $k_B T$. Thus the step free energy $\beta(\phi)$ per unit length can be estimated as

$$\beta(\phi) = \beta_0 + c k_B T p_K \frac{a}{\ell^2} \approx \beta_0 + \frac{c k_B T}{a} \frac{\exp(-J_x/k_B T)}{1 + 2\exp(-J_x/k_B T)} \phi^2 \equiv \beta_0 + \beta_2 \phi^2, \quad (11.4)$$

where β_0 is the step free energy of an isolated step obtained in the previous section, and the second term represents the contribution from the step interaction with c being some constant. The coefficient β_2 is inversely proportional to the stiffness of an isolated step as $\beta_2 \propto (k_B T)^2/\tilde{\beta}(0)$. It reflects the fact that the step interaction is brought about by the step fluctuation.

Exercise:(Gruber-Mullins model [72]) In order to treat the entropy reduction by the step confinement more quantitatively, we consider a step confined between two walls. For simplicity we consider every length in unit of lattice constant a here. The wall is located at $y = 0$ and $y = 2\ell$, so that the step height $y(i)$ at the site with the x-coordinate i can take values only from 1 to $2\ell - 1$. The system has the interaction Hamiltonian (10.1). The partition function is transformed in terms of the transfer matrix $\hat{T}$ as

$$\begin{aligned} Z &= \exp\left(-\frac{L J_y}{k_B T}\right) \sum_{y(1)=1}^{2\ell-1} \cdots \sum_{y(L)=1}^{2\ell-1} \hat{T}(y(1), y(2)) \cdots \hat{T}(y(L), y(1)) \\ &= \exp\left(-\frac{L J_y}{k_B T}\right) \mathrm{Tr} \hat{T}^L. \end{aligned} \quad (11.5)$$

Here Tr means the trace of the matrix, and $\hat{T}$ is the symmetric $(2\ell - 1) \times (2\ell - 1)$ transfer matrix.

(1) Show that $\hat{T}$ is explicitly written as

$$\hat{T}(h, h') = \begin{pmatrix} 1 & B & 0 & 0 & \cdot & 0 \\ B & 1 & B & 0 & \cdot & 0 \\ 0 & B & 1 & B & \cdot & 0 \\ \cdot & \cdot & \cdot & \cdot & \cdot & \cdot \\ 0 & \cdot & \cdot & 0 & B & 1 \end{pmatrix} \quad (11.6)$$

by using the abbreviation of the Boltzmann factor as $B = e^{-J_x/k_B T}$.

(2) If one knows the eigenvalues μ_k $(k = 1 \sim 2\ell - 1)$ of $\hat{T}$, the trace is easily calculated as

$$\mathrm{Tr}\hat{T}^L = \sum_{k=1}^{2\ell-1} \mu_k^L. \qquad (11.7)$$

For a long step, $L \to \infty$, only the largest eigenvalue μ_1 gives the dominant contribution, and the free energy F is obtained by

$$F = -k_B T \ln Z = L\left[J_y - k_B T \ln \mu_1\right]. \qquad (11.8)$$

By denoting the eigenvectors as $\psi = (\psi(1), \psi(2), \cdots, \psi(2\ell - 1))$, explicitly write down the eigenvalue equation $\hat{T}\psi = \mu\psi$.

(3) Since the eigenvalue equation corresponds to the wave equation with a fixed boundary condition, the eigenvector ψ_k is written as

$$\psi_k(m) = \sin\frac{\pi k m}{2\ell} \qquad (k = 1 \cdots 2\ell - 1). \qquad (11.9)$$

Find the corresponding eigenvalue μ_k.

(4) Find the step free energy per length $\beta(\phi)$ of a vicinal face with an angle (11.1), and show that for small ϕ or large ℓ, $\beta(\phi)$ can be expanded as

$$\beta(\phi) = \beta_0 + \beta_2\phi^2. \qquad (11.10)$$

Answer:

(2) For $m = 1 \cdots (2\ell - 1)$, eigenvalue equations are written as

$$B\left[\psi(m+1) - 2\psi(m) + \psi(m-1)\right] + (2B + 1 - \mu)\psi(m) = 0, \qquad (11.11)$$

and $\psi(0) = \psi(2\ell) = 0$.

(3) From the simple trigonometry, one obtains the eigenvalues

$$\mu_k = 1 + 2B\cos\frac{\pi k}{2\ell} \qquad (11.12)$$

for $k = 1 \sim 2\ell - 1$.

(4) Step free energy is calculated as

$$\begin{aligned}
\beta(\phi) &= \frac{F}{L} = -\frac{k_B T \ln Z}{L} = J_y - k_B T \ln\left[1 + 2\exp\left(-\frac{J_x}{k_B T}\right)\cos\frac{\pi}{2\ell}\right] \\
&\approx J_y - k_B T \ln\left[1 + 2\exp(-\frac{J_x}{k_B T})\right] + \frac{2k_B T \exp(-J_x/k_B T)}{1 + 2\exp(-J_x/k_B T)}\left(1 - \cos\frac{\pi}{2\ell}\right) \\
&= \beta_0 + \beta_2\phi^2,
\end{aligned} \qquad (11.13)$$

where β_0 is given in Eq.(10.3) and

$$\beta_2 = \frac{\pi^2 k_B T}{8a} \frac{2\exp(-J_x/k_B T)}{1 + 2\exp(-J_x/k_B T)} = \frac{\pi^2(k_B T)^2}{8a^2\tilde{\beta}(0)}. \qquad (11.14)$$

In the last equation, the lattice parameter a is explicitly written again. If the neighboring steps are allowed to fluctuate, the exact calculation shows [3] that

$$\beta_2 = \frac{\pi^2(k_B T)^2}{6a^2\tilde{\beta}(0)}. \qquad (11.15)$$

Exercise 2:

(1) Show that when the height of the vicinal face deviates $\bar{z}(x,y)$ from the mean position, the surface free energy is described by [146]

$$\mathcal{H} = \int dx\, dy \left[\frac{1}{2}\tilde{\gamma}_\perp \left(\frac{\partial \bar{z}}{\partial x} \right)^2 + \frac{1}{2}\tilde{\gamma}_\parallel \left(\frac{\partial \bar{z}}{\partial y} \right)^2 \right], \qquad (11.16)$$

with two surface stiffness constants

$$\tilde{\gamma}_\perp = \tilde{\beta}\frac{\ell_0}{a^2} \qquad (11.17)$$

and

$$\tilde{\gamma}_\parallel = \frac{(\pi kT)^2}{\tilde{\beta}\ell_0 a^2}. \qquad (11.18)$$

Here $\tilde{\beta}$ is the stiffness of the steps running on average in x direction with the separation ℓ_0 in y direction. a is the jump in the terrace height at the step.

(2) Show that the width of a single step fluctuation is proportional to the logarithm of the system size L as

$$w^2 = \langle h^2 \rangle = \frac{\ell_0^2}{2\pi^2}\ln L, \qquad (11.19)$$

by using the surface free energy (11.16),

Answer:

(1) When steps are straight and aligned regularly as in Fig.11.2(a), the vicinal face $\bar{z}(x,y) = 0$ is tilted from the terrace face by an angle $\phi_0 = \arctan(a/\ell_0)$. The step fluctuation induces height fluctuation, and we consider two independent modes of step fluctuation as follows.

Let every step deforms locally at a position x in y direction by $h(x)$, while keeping the step separation ℓ_0 constant, as shown in Fig.11.2(b). This step

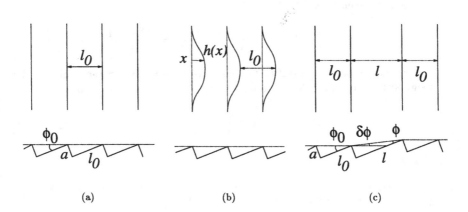

Figure 11.2: (a) Flat vicinal face, (b) step bending deformation, and (c) step dilatation.

deformation corresponds to the height increment $\bar{z}(x, y) = h(x)a/\ell_0$. For a step segment of length dx, the energy increase is given by

$$\frac{1}{2}\tilde{\beta}\left(\frac{dh}{dx}\right)^2 dx \qquad (11.20)$$

with the step stiffness $\tilde{\beta}$. In the interval dy in y direction, there are dy/ℓ_0 steps and the energy increase in an area element $dxdy$ is obtained as

$$\Delta E_\perp = \frac{1}{2}\tilde{\beta}\left(\frac{\ell_0}{a}\right)^2\left(\frac{\partial\bar{z}}{\partial x}\right)^2 dx\frac{dy}{\ell_0} = \frac{1}{2}\tilde{\gamma}_\perp\left(\frac{\partial\bar{z}}{\partial x}\right)^2 dxdy, \qquad (11.21)$$

where

$$\tilde{\gamma}_\perp = \tilde{\beta}\frac{\ell_0}{a^2} \approx \frac{\tilde{\beta}}{a\phi_0}. \qquad (11.22)$$

The second mode of step deformation is the compression or dilatation of the step train, as shown in Fig.11.2(c). When the straight steps shift in y direction and one of the terrace width increases from ℓ_0 to ℓ, the tilt angle of the vicinal face there in y direction changes by an angle $\delta\phi = \arctan(\partial\bar{z}/\partial y)$. This portion is inclined from the terrace face by an angle $\phi = \arctan(a/\ell)$. Three angles ϕ, ϕ_0, $\delta\phi$ are related as $\phi_0 = \phi + \delta\phi$, and for a small tilt relations are reduced to

$$\frac{a}{\ell} = \frac{a}{\ell_0} - \frac{\partial\bar{z}}{\partial y}. \qquad (11.23)$$

Now we calculate the energy change caused by the step compression. As is explained in the main text of the section, the step collision induces the energy increase. For a step separation ℓ_0, the collision takes place along the step for every length of L_c determined from the relation $\ell_0^2 \approx k_B T L_c / \tilde{\beta}$. The energy increase by each collision is of the thermal energy $k_B T$. In an area $dxdy$, there are dy/ℓ_0 steps and each step makes dx/L_c collisions, and thus the total energy increase is proportional to

$$k_B T \frac{dx}{L_c} \frac{dy}{\ell_0} = \frac{(k_B T)^2}{\tilde{\beta} \ell_0^3} dxdy. \tag{11.24}$$

The precise calculation [3, 161] shows that the energy increase is $\pi^2/6$ times larger than the above estimation. Therefore, the terrace width variation leads the energy increase of

$$\begin{aligned}
\Delta E_\parallel &= \frac{\pi^2}{6} \frac{(k_B T)^2}{\tilde{\beta}} \left(\frac{1}{\ell^3} - \frac{1}{\ell_0^3} \right) dxdy = \frac{\pi^2}{6} \frac{(k_B T)^2}{\tilde{\beta}} \left[\left(\frac{1}{\ell_0} - \frac{1}{a} \frac{\partial \bar{z}}{\partial y} \right)^3 - \frac{1}{\ell_0^3} \right] dxdy \\
&= \frac{(\pi k_B T)^2}{6\tilde{\beta}} \left[-\frac{3}{\ell_0^2 a} \frac{\partial \bar{z}}{\partial y} + \frac{3}{\ell_0 a^2} \left(\frac{\partial \bar{z}}{\partial y} \right)^2 - \frac{1}{a^3} \left(\frac{\partial \bar{z}}{\partial y} \right)^3 \right] dxdy. \tag{11.25}
\end{aligned}$$

Since the average orientation of the whole vicinal face is fixed,

$$\int_0^{L_y} \frac{\partial \bar{z}}{\partial y} dy = \bar{z}(L_y) - \bar{z}(0) = 0 \tag{11.26}$$

and the linear term in the slope $\partial \bar{z}/\partial y$ is irrelevant in the free energy (11.25). By neglecting the third order term of the slope in (11.25), we get

$$\Delta E_\parallel \approx \frac{1}{2} \tilde{\gamma}_\parallel \left(\frac{\partial \bar{z}}{\partial y} \right)^2 dxdy \tag{11.27}$$

with

$$\tilde{\gamma}_\parallel = \frac{(\pi k T)^2}{\tilde{\beta} \ell_0 a^2} \approx \frac{(\pi k_B T)^2}{\tilde{\beta} a^3} \phi_0. \tag{11.28}$$

By summing up two contributions, (11.21) and (11.27), we get the Hamiltonian (11.16). Note the strong anisotropy of the surface stiffness. For a small tilt angle ϕ_0 $\tilde{\gamma}_\perp$ diverges whereas $\tilde{\gamma}_\parallel$ vanishes. This drastic difference of two surface stiffness constants is observed recently in He [158]. The product $\tilde{\gamma}_\perp \tilde{\gamma}_\parallel$ remains finite though:

$$\tilde{\gamma}_\perp \tilde{\gamma}_\parallel = \frac{\tilde{\beta}}{a\phi_0} \frac{(\pi k_B T)^2 \phi_0}{\tilde{\beta} a^3} = \frac{(\pi k_B T)^2}{a^4}. \tag{11.29}$$

(2) From the free energy (11.16), the Fourier-transformed height $\bar{z}(q_x, q_y)$ has the correlation

$$\langle |\bar{z}(q_x, q_y)|^2 \rangle = \frac{k_B T}{\tilde{\gamma}_\perp q_x^2 + \tilde{\gamma}_\parallel q_y^2}. \tag{11.30}$$

The height-difference correlation function of the vicinal face is then calculated as

$$\langle(\bar{z}(r) - \bar{z}(0))^2\rangle = 2\int\frac{d^2q}{(2\pi)^2}\langle|\bar{z}(q_x, q_y)|^2\rangle(1 - \cos qr) = \frac{k_BT}{\pi\sqrt{\tilde{\gamma}_\perp\tilde{\gamma}_\parallel}}\ln r = \frac{a^2}{\pi^2}\ln r.$$
(11.31)

Since the height $\bar{z}(x, 0)$ is related to the step deformation $h(x)$ as $\bar{z}(x, 0) = h(x)a/\ell_0$, the correlation function of the step deformation is shown to diverge for large separation along the step as

$$\langle(h(x) - h(0))^2\rangle = \frac{\ell_0^2}{a^2}\langle(z(x, 0) - z(0, 0))^2\rangle = \frac{\ell_0^2}{\pi^2}\ln x.$$
(11.32)

One can also calculated the height fluctuation at a point r for a finite system of size L^2 as

$$\langle\bar{z}(r)^2\rangle = \frac{k_BT}{2\pi\sqrt{\tilde{\gamma}_\perp\tilde{\gamma}_\parallel}}\ln L = \frac{a^2}{2\pi^2}\ln L.$$
(11.33)

This result is translated in the step width as

$$w^2 = \langle h^2\rangle = \frac{\ell_0^2}{a^2}\frac{k_BT}{2\pi\sqrt{\tilde{\gamma}_\perp\tilde{\gamma}_\parallel}}\ln L = \frac{\ell_0^2}{2\pi^2}\ln L.$$
(11.34)

The width of a step in a step train diverges weaker than that of an isolated step due to the confinement effect of neighboring steps.

12   Faceting Transition

We now study the effect of surface roughening on the crystal shape in equilibrium. The equilibrium shape is determined by Andreev's formulation as   Eqs.(6.8) and (6.11). Therefore, once the free energy f is known, the equilibrium shape can be calculated.

Below the roughening temperature T_R, the steps on a vicinal surface with an inclination angle ϕ in y-direction (Fig.11.1a) has the step free energy (11.4) or

$$\beta(\phi) = \beta_0 + \beta_2\phi^2.$$
(12.1)

Then the surface free energy per xy projected unit area, (11.2), is represented as

$$f(\phi) = \frac{\gamma(\phi)}{\cos\phi} = \gamma_0 + \frac{\beta(\phi)|\phi|}{a} = \gamma_0 + \frac{\beta_0|\phi|}{a} + \frac{\beta_2|\phi|^3}{a}.$$
(12.2)

Here we have used the symmetry consideration that the surface with positive and negative tilt have the same density of steps and should have the same f. Thus f has a cusp singularity at $\phi = 0$ as shown in Fig.12.1a. The vicinal face with a slope

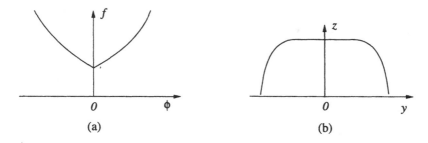

Figure 12.1: (a) Surface free energy f per xy projected area, and (b) the faceted equilibrium shape near $y = 0$ below the roughening temperature T_R.

$p_y = \partial z/\partial y = \tan\phi \approx \phi$ is located at a distance y apart from the center, where y is given by Eq. (6.8) as

$$-\lambda y = \frac{\partial f}{\partial \phi} = \frac{1}{a}\left(\beta_0 + 3\beta_2\phi^2\right)\mathrm{sgn}(\phi). \tag{12.3}$$

Here $\mathrm{sgn}(\phi)$ represents the sign of the slope ϕ. Since λ, β_0 and β_2 are positive for $T < T_R$, the slope is positive for $y < 0$ and negative for $y > 0$. Let us consider the region of $y > 0$ and $\phi < 0$. (See Fig.12.1.) Then Eq.(12.3) is solved for ϕ as

$$\phi = -\sqrt{\frac{\lambda a}{3\beta_2}(y - y_0)} \tag{12.4}$$

for y larger than $y_0 \equiv \beta_0/\lambda a$. The height z is determined from Eq.(6.11) as

$$\lambda z = f + \lambda y p_y = \gamma_0 - 2a^{-1}\beta_2|\phi|^3. \tag{12.5}$$

The crystal shape is then written as

$$z - z_0 = -\frac{2\beta_2}{\lambda a}\left(\frac{\lambda a}{3\beta_2}\right)^{3/2}(y - y_0)^{3/2} \tag{12.6}$$

for $y > y_0$. Here $z_0 = \gamma_0/\lambda$. For $y < -y_0$, ϕ is positive and the profile is similar to Eq.(12.6) with only a replacement of y to $-y$. Therefore, between $-y_0$ and y_0 the shape consists of a facet face, and then it continues to the smooth curved faces with the exponent $3/2$, as shown in Fig.12.1b [160]. The facet size $2y_0$ is proportional to the step free energy density, β_0. Near the roughening point T_R, β_0 and thus y_0 vanish smoothly as $y_0 \sim \exp[-c/\sqrt{T_R - T}]$. This extinction of the facet face is called the faceting phase transition, and is equivalent to the roughening transition of the face.

There are experimental studies on faceting transitions of ^{4}He crystals in superfluid liquid. The transition temperature is determined to be $T_R = 1.3$K for (0001) face. The roughening temperatures of other faces are also determined [68].

Part III

Kinetics-Limited Growth

In this part, the growth laws governed by the surface kinetics on a flat crystal surfaces are studied. When the crystal grows with a flat surface, the growth is controlled by the incorporation of atoms to steps and kinks provided by the two-dimensional nuclei [12] or screw dislocations [62]. Since the step density depends on the growth condition, the growth law is different from the ideal ones.

13   Step Advancement by Surface Diffusion

We consider the crystal growth from the vapor phase at a relatively low temperature where a crystal surface is flat. Atoms deposited on the surface diffuse around on the surface before they reach steps. The advancement of a step in the surface diffusion field of adsorbed atoms is studied systematically by Burton, Cabrera and Frank [43].

We assume that the adsorption site on a crystal surface forms a square lattice with a lattice constant a. From the ambient gas phase f atoms deposit on a surface per unit area in a unit time interval. Therefore, on a lattice site (x, y) with atomic area $\Omega_2 = a^2$, $f\Omega_2\Delta t$ atoms are deposited in a time interval Δt. The adsorbed atom vibrates with a frequency ν of the lattice vibration around the averaged adsorption site, and tries evaporation or surface diffusion.   Due to the coupling with the underlying crystal atoms, the adatom has an energy lower than in the free state by E_{ad}, and the evaporation takes place with the probability $\exp(-E_{ad}/k_BT)$ during ν trials per unit time. The lifetime until the evaporation is estimated as

$$\tau = \nu^{-1}\exp(E_{ad}/k_BT). \tag{13.1}$$

By denoting the adatom density as $c(x, y)$, the probability that a lattice site (x, y) is occupied by an adatom at a time t is $c(x, y; t)\Omega_2$. The number of adatoms evaporating within a time interval Δt is given by $c(x, y; t)\Omega_2 \cdot \Delta t/\tau$.

On the other hand, an adatom hops to one of the neighboring sites during its lifetime. For the random walk of the adatom from one adsorption site to the other it has to cross over the activation energy or energy barrier E_{sd}, and the probability of hopping to the neighboring site is $w = \nu\exp(-E_{sd}/k_BT)$. The number of atoms coming from the four neighboring sites during the time Δt is $w\Delta t\Omega_2[c(x + a, y; t) + c(x - a, y; t) + c(x, y + a; t) + c(x, y - a; t)]$, while the number of atoms escaping from the site (x, y) is $w\Delta t\Omega_2 4c(x, y; t)$. Combining all these processes, the variation of the number of averaged atoms at the site (x, y) during the time interval Δt is written as

$$c(x, y; t + \Delta t)\Omega_2 = c(x, y; t)\Omega_2 + f\Omega_2\Delta t - c(x, y; t)\Omega_2\frac{\Delta t}{\tau}$$

$$+ w\Omega_2\Delta t[c(x+a, y; t) + c(x-a, y; t) + c(x, y+a; t) + c(x, y-a; t) - 4c(x, y; t)]. \tag{13.2}$$

After dividing Eq.(13.2) by $\Omega_2 \Delta t$, take the continuum limit $\Delta t \to 0$ and $a \to 0$ and leave up to the second derivative of the space. Then one obtains the diffusion equation including the contribution of deposition flux f and the desorption lifetime as

$$\frac{\partial c(x, y; t)}{\partial t} = D_s \nabla^2 c + f - \frac{c}{\tau}. \tag{13.3}$$

Here

$$D_s = wa^2 = \nu a^2 \exp(-E_{sd}/k_B T) \tag{13.4}$$

is the surface diffusivity.

How far can the adsorbed atom meander while it resides on the surface? During the time t, an adatom makes wt jumps on average. Since it can jump to four nearest neighbors with the same probability, the average position $\langle \mathbf{r}(t) \rangle$ remains the same with the original one, $\langle \mathbf{r}(0) \rangle$. But the deviation $\langle (\mathbf{r}(t) - \mathbf{r}(0))^2 \rangle$ increases in proportional to the number of jumps. Since it hops a distance a in a single jump, the deviation is written as $\langle (\mathbf{r}(t) - \mathbf{r}(0))^2 \rangle = wta^2 = D_s t$. Therefore, during its lifetime τ on the surface, an adatom meanders in the range

$$x_s = \sqrt{D_s \tau}, \tag{13.5}$$

which is called mean displacement of the adsorbed atom or the surface diffusion length.

During the surface diffusion an adatom reaches a step and then crystallizes by being incorporated in the kink sites. Let us consider the advancement of an isolated step running on average in the x-direction. Usually the step advances so slowly compared to the relaxation of the adatom density c that the stationary distribution of c can be realized: The time derivative in Eq.(13.3) is disregarded. If the step is straight in the x-direction, the concentration variation takes place only in y-direction, and the diffusion equation is simplified as

$$D_s \frac{d^2 c}{dy^2} = \frac{c - c_\infty}{\tau}, \tag{13.6}$$

where $c_\infty = f\tau$ is the concentration far from the step. The solution is easily obtained as

$$c - c_\infty = A \exp\left(-\frac{|y|}{x_s}\right) \tag{13.7}$$

with the surface diffusion length x_s. Since the step is rough at finite temperatures as shown in section 10, the kinetics of adatom incorporation is expected to be very fast. In the extreme case, the step can grow with only an infinitesimal driving as $c(0) = c_{eq}$. From this local equilibrium boundary condition at the step, the integration constant A is determined to be $A = c_{eq} - c_\infty$, and the diffusion field is obtained as

$$c(x, y; t) = c_\infty - (c_\infty - c_{eq}) \exp\left(-\frac{|y|}{x_s}\right), \tag{13.8}$$

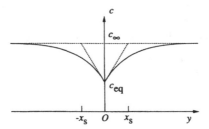

Figure 13.1: Concentration distribution around the step at $y = 0$.

as is depicted in Fig.13.1.

The net number of atoms impinging to the step at $y = 0$ from the right ($y > 0$) is $w\Omega_2\left[c(x, a; t) - c(x, 0; t)\right] = wa^3 \, \partial c(x, y)/\partial y|_{y=+0}$, and that from the left ($y < 0$) is $w\Omega_2\left[c(x, -a; t) - c(x, 0; t)\right] = -wa^3 \, \partial c(x, y)/\partial y|_{y=-0}$. Here $y = +0(-0)$ means in front (in the back) of the step. Since these deposited adatoms crystallize at a kink site on a step, and step advances a distance a by each crystallization, the advancement rate of a straight step is obtained as

$$v_0 = a\left[wa^3\left(\frac{\partial c}{\partial y}\right)_{+0} - wa^3\left(\frac{\partial c}{\partial y}\right)_{-0}\right] = D_s\Omega_2\left[\left(\frac{\partial c}{\partial y}\right)_{+0} - \left(\frac{\partial c}{\partial y}\right)_{-0}\right]. \qquad (13.9)$$

By inserting the density profile (13.8), the step velocity is obtained as

$$v_0 = (c_\infty - c_{eq})\Omega_2\frac{2D_s}{x_s} = (f - f_{eq})\Omega_2 \cdot 2x_s. \qquad (13.10)$$

Here $f_{eq} = c_{eq}/\tau$ is the equilibrium deposition flux. The result (13.10) can be interpreted as follows: The adsorbed atoms landed on a surface within a distance of their mean displacement x_s from a step will eventually meet the step during their lifetime, and be incorporated to it. Therefore, $f \times 2x_s a$ atoms impinged in a surface region of a step length a within a distance x_s from the step will crystallize at a step in a unit time. As an inverse process, $f_{eq} \times 2x_s a$ atoms melt back from the step onto the surface. Since each crystallized atom pushes the step ahead by a distance a , the advance rate is given as $v_0 = (f2x_s a - f_{eq}2x_s a) \times a$ in agreement with the result of the detailed calculation (13.10).

So far we considered an isolated step on a singular flat surface. When the surface is tilted with a small angle, there are many steps arranged periodically: The vicinal surface consists of a step train. When the separation ℓ between periodic steps is smaller than the diffusion length x_s, the diffusion field of neighboring steps overlaps and competes. A single step gets a contribution only from the small range around itself with a width ℓ. Therefore the advance velocity will decrease to $v = (f - f_{eq})a^2\ell = v_0 \times (\ell/2x_s)$. Of course, when the step separation ℓ is wider than the surface diffusion length x_s, each step advances with a velocity v_0 as if it is an isolated single step. In a more precise treatment, the solution of the diffusion equation (13.3) is written as $c(y) - c_\infty = A \exp(-y/x_s) + B \exp(y/x_s)$ within a region $0 < y < \ell$. By taking into account the local equilibrium boundary conditions at neighboring steps, $c(y = 0) = c(y = \ell) = c_{eq}$, the integration constants A and B are determined, and the concentration is obtained as

$$c(x, y; t) = c_\infty - (c_\infty - c_{eq}) \frac{\cosh\left[(2y - \ell)/2x_s\right]}{\cosh(\ell/2x_s)}. \qquad (13.11)$$

From the material conservation (13.9), the advance velocity of a step is given as

$$v = (f - f_{eq})a^2 2x_s \tanh\left(\frac{\ell}{2x_s}\right) = v_0 \tanh\left(\frac{\ell}{2x_s}\right). \qquad (13.12)$$

If the step separation ℓ is much larger than the diffusion length x_s, the advance rate v is equal to the isolated step v_0 in (13.10). If ℓ is smaller than x_s, then the approximation $\tanh(\ell/2x_s) \approx \ell/2x_s$ holds and the velocity is approximated as $v \approx (f - f_{eq})a^2\ell$, as already stated.

Exercise: Derive Eq.(13.12).
Answer: By imposing a local equilibrium boundary conditions at $y = 0$ and $y = \ell$, one gets

$$\begin{aligned} c(0) - c_\infty &= A + B = c_{eq} - c_\infty \\ c(\ell) - c_\infty &= Ae^{-\ell/x_s} + Be^{\ell/x_s} = c_{eq} - c_\infty, \end{aligned} \qquad (13.13)$$

and the integral constants are obtained as

$$A = (c_{eq} - c_\infty)\frac{e^{\ell/2x_s}}{2\cosh(\ell/2x_s)} \qquad (13.14)$$

and

$$B = (c_{eq} - c_\infty)\frac{e^{-\ell/2x_s}}{2\cosh(\ell/2x_s)}. \qquad (13.15)$$

By putting these results, one gets (13.11). From Eq.(13.9), we get the velocity (13.12).

14 Advancement of a Circular Step

When a two-dimensional crystal nucleus grows in a diffusion field, the encircling step expands and the total step free energy increases. In order for the nucleus to achieve further expansion, this energy cost has to be overcome. How will then the step advancement rate be modified from the straight step?

The chemical potential of the bulk crystal μ_S agrees with that of the adsorbed atoms with the density c_{eq} in equilibrium: $\mu_S = \mu_{ad}(c_{eq})$. If a circular nucleus with a radius ρ is in equilibrium with adatoms, the radius corresponds to the critical one since it does not grow or shrink. Then the chemical potential $\mu_{ad}(c_{eq}(\rho))$ for adatom with density $c_{eq}(\rho)$ should satisfy the critical condition (6.14): $\Delta\mu = \mu_{ad}(c_{eq}(\rho)) - \mu_S = \beta\Omega_2/\rho$. Here the step free energy β is assumed isotropic. Since the ideal solution approximation (5.2) is applicable for a dilute adatom system, the chemical potential difference is written in terms of adatom density as

$$c_{eq}(\rho) = c_{eq} \exp\left(\frac{\Delta\mu}{k_B T}\right) = c_{eq} \exp\left(\frac{\beta\Omega_2}{\rho k_B T}\right) \approx c_{eq}\left(1 + \frac{\beta\Omega_2}{\rho k_B T}\right). \tag{14.1}$$

The last approximation is valid for a step with a small curvature $\kappa = 1/\rho$. The curved step needs a higher equilibrium concentration than the straight step does in order to remain in equilibrium. The effect is called the Gibbs-Thomson effect.

The adatom density far from the nucleus is determined by the deposition flux as $c_\infty = f\tau$. When the density c_∞ is higher than the equilibrium value $c_{eq}(\rho)$, the circular nucleus with a radius ρ grows. The density distribution around the circular nucleus is expected to be symmetric and to depend only on the radius r. The diffusion equation is simplified as

$$\frac{\partial c(r;t)}{\partial t} = D_s\left(\frac{\partial^2 c}{\partial r^2} + \frac{1}{r}\frac{\partial c}{\partial r}\right) - \frac{c - c_\infty}{\tau}. \tag{14.2}$$

For the slow growth, the stationary approximation $\partial c/\partial t = 0$ is expected to hold. The exact solution of c is described in Appendix A14. Here I derive the circular step advancement in approximation.

The radius of the nucleus ρ is assumed large. Since the step advancement is controlled by the diffusion field around the step, the radial variable r in Eq.(14.2) is also large, and the second term, $r^{-1}\partial c/\partial r$, may be neglected compared to the first term, $\partial^2 c/\partial r^2$. The stationary density distribution is obtained as

$$c(r) - c_\infty = \begin{cases} A_+ e^{-(r-\rho)/x_s} & \text{for } r > \rho \\ A_- e^{(r-\rho)/x_s} & \text{for } r < \rho \end{cases} \tag{14.3}$$

with $x_s = \sqrt{D_s\tau}$ being the surface diffusion length defined in Eq.(13.5). The adatom density relaxes to c_∞ far outside ($r \gg \rho$) and far inside ($r \ll \rho$) of the cluster. By imposing the local equilibrium condition $c(\rho) = c_{eq}(\rho)$ at the circular step, the integral constant $A_\pm$ is determined as $A_\pm = c_{eq}(\rho) - c_\infty$. Then the step advancement velocity in the radial direction is obtained as

$$\begin{aligned} v(\rho) &= D_s a^2 \left(\left.\frac{\partial c}{\partial r}\right|_{\rho+0} - \left.\frac{\partial c}{\partial r}\right|_{\rho-0}\right) = \frac{2D_s a^2}{x_s}(c_\infty - c_{eq}(\rho)) \\ &= v_0\left(1 - \frac{\exp(\beta\Omega_2/\rho k_B T) - 1}{c_\infty/c_{eq} - 1}\right) \approx v_0\left(1 - \frac{\rho_c}{\rho}\right), \end{aligned} \tag{14.4}$$

where $v_0 = 2D_s\Omega_2(c_\infty - c_{eq})/x_s$ is the velocity of the straight step (13.10), and

$$\rho_c = \frac{\beta\Omega_2}{k_BT(c_\infty/c_{eq} - 1)} = \frac{\beta\Omega_2}{k_BT(f/f_{eq} - 1)} \tag{14.5}$$

is the critical radius corresponding to the deposition flux f or Eq.(6.14). Therefore, the nucleus with critical radius ρ_c does not grow or shrink, the one with radius $\rho > \rho_c$ grows, and the one with $\rho < \rho_c$ shrinks. The correct calculation in Appendix A14 leads to the same conclusion.

15 Growth Rate by Spiral Growth Mechanism

The growth rate V of a crystal by spiral growth mechanism is now ready to be calculated. The step is running out from the center of the screw dislocation. As the crystallization proceeds, the initially straight step turns round the dislocation, and the curvature at the center increases, as shown in Fig.10.1. But it should be smaller than the critical one, $\kappa_c = 1/\rho_c$. Otherwise, the crystal melts back due to the Gibbs-Thomson effect. Therefore, the curvature at the center is just the critical one in the steady state. The simplest form of the spiral is that of Archimedes, represented in the two-dimensional polar coordinate (r, ϕ) as $r = a\phi$ with a to be determined. As is explained in the last paragraph of Appendix A15, the curvature at $\phi \to 0$ is calculated to be $\kappa = 2/a$. Therefore, the parameter a should be $2\rho_c$ in order to satisfy the boundary condition. The step separation ℓ for large r is given as the radius difference by one turn, or $\Delta\phi = 2\pi$ rotation as

$$\ell = a \cdot 2\pi = 4\pi\rho_c. \tag{15.1}$$

The more sophisticated calculation is described in Appendix A15 [44]. It shows that the asymptotic step separation for large r is

$$\ell \approx 19\rho_c. \tag{15.2}$$

In both cases, step separation ℓ increases in inversely proportional to the driving force, $\ell \sim \Delta\mu^{-1}$. Since the advancement velocity $v(\ell)$ of a step train is given by Eq.(13.2), the time T required for the step to wind up one turn is given by $\ell/v(\ell)$. During this time T the crystal surface grows a height a in the normal direction, and its growth rate of the crystal is then given as

$$V = v(\ell)\frac{a}{\ell} = a^3(f - f_{eq})\frac{2x_s}{\ell}\tanh\left(\frac{\ell}{2x_s}\right), \tag{15.3}$$

and the dependence on the driving force $\Delta\mu$ is shown in Fig. 15.1.

For a low vapor pressure p, the step separation ℓ is larger than the surface diffusion length x_s, and $\tanh(\ell/2x_s) \approx 1$. The growth rate is approximately given by $V =$

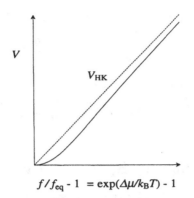

Figure 15.1: Growth velocity V versus driving force $\sigma = \exp(\Delta\mu/k_BT) - 1$ for the spiral growth.

$a^3(f - f_{\mathrm{eq}})2x_\mathrm{s}/\ell$ proportional to the square of the driving force $(p - p_{\mathrm{eq}})^2 \sim \Delta\mu^2$. Thus, at low vapor pressure growth follows the parabolic law, different from the linear growth law of Hertz-Knudsen. When the vapor pressure p increases, the driving force for the crystal growth $\Delta\mu$ becomes large, and the critical radius ρ_c and the step separation ℓ becomes small. For $\ell \ll x_\mathrm{s}$, the approximation $\tanh(\ell/2x_\mathrm{s}) \approx \ell/2x_\mathrm{s}$ holds, and the growth rate reduces to the Hertz-Knudsen formula given in Eq.(4.4): $V_{\mathrm{HK}} = a^3(f - f_{\mathrm{eq}}) = a^3(p - p_{\mathrm{eq}})/\sqrt{2\pi m k_B T}$: There are many steps and kinks on the surface, and the growth rate V reduces to that of a rough surface.

16   Two-Dimensional Nucleation and Growth

Spiral mechanism governs the growth of an imperfect crystal under a small driving force. When the supersaturation increases, nucleation becomes important. Also for a perfect crystal without dislocation, nucleation controls the crystal growth. In this section the growth rate of a crystal by nucleation mechanism is studied [12, 80, 69].

The formation of a two-dimensional (2D) nucleus increases the free energy as

$$- \Delta\mu \frac{\pi\rho^2}{\Omega_2} + 2\pi\rho\beta = -\Delta\mu \cdot n + \beta\sqrt{4\pi n\Omega_2} \equiv G(n), \tag{16.1}$$

where ρ is the radius of the 2D nucleus, $\Delta\mu = \mu_\mathrm{G} - \mu_\mathrm{S}$ the chemical potential gain by the crystallization, Ω_2 the area of a nucleating atom, β the step free energy per length and assumed to be isotropic here. There are $n = \pi\rho^2/\Omega_2$ atoms in the circular nucleus of radius ρ. $G(n)$ has a maximum as shown in Fig.16.1 at the critical nucleus size n_c:

$$n_\mathrm{c} = \frac{\pi\rho_\mathrm{c}^2}{\Omega_2} = \frac{\pi\beta^2\Omega_2}{(\Delta\mu)^2}. \tag{16.2}$$

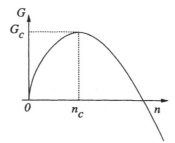

Figure 16.1: Free energy cost $G(n)$ of the formation of the two-dimensional circular nucleus with the size n.

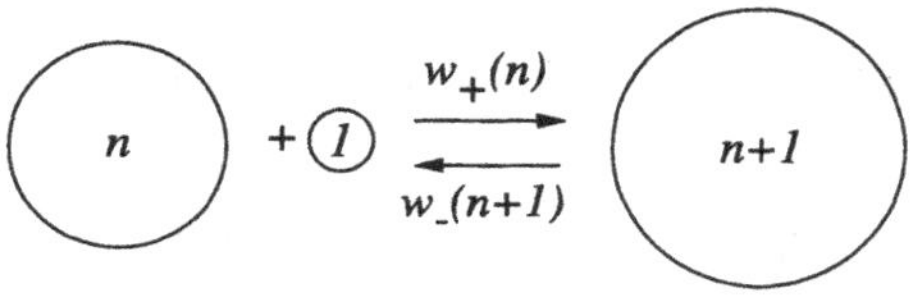

Figure 16.2: Variation of the cluster size n by capturing or by evaporating a single adatom.

The maximum value of the free energy barrier is

$$G_c = G(n_c) = \Delta\mu n_c = \frac{\pi\beta^2\Omega_2}{\Delta\mu}. \tag{16.3}$$

Let us consider the distribution $C(n, t)$ of the cluster of a size n at time t. In equilibrium thermal fluctuation allows the formation of nucleus of a cluster size n with the Boltzmann weight $C^0(n) = c_\infty \exp(-G(n)/k_BT)$, where c_∞ is the density of isolated adatoms: $c_\infty = f\tau$. Note that the equilibrium density of infinitely large nucleus is infinite. During the steady growth of a crystal, in fact, a large nucleus disappears by the completion of a layer and the density of an infinitely large cluster in a steady state should be zero $C(n \to \infty) = 0$.

The size variation of the nucleus is assumed to take place by capturing or evaporating a single adatom as shown in Fig.16.2. The rate equation is written as

$$\frac{\partial C(n)}{\partial t} = w_+(n-1)C(n-1) + w_-(n+1)C(n+1) - w_+(n)C(n) - w_-(n)C(n), \tag{16.4}$$

where $w_+(n)$ is the rate of capture of a single adatom by the nucleus with the size n and $w_-(n)$ is the rate of evaporation of an atom from the nucleus of size n. The detailed balance condition

$$w_+(n)C^0(n) = w_-(n+1)C^0(n+1) \tag{16.5}$$

confirms the approach to equilibrium. Then the ratio of the capture and evaporation rates is obtained as

$$\frac{w_+(n)}{w_-(n+1)} = \frac{C^0(n+1)}{C^0(n)} = \frac{\exp[-G(n+1)/k_BT]}{\exp[-G(n)/k_BT]}. \tag{16.6}$$

In a steady state, $\partial C(n)/\partial t = 0$ but there is a net flow J of the cluster growth supported by the constant supply of particles from the smallest size and their annihilation at the largest size. J is then independent of the cluster size as

$$J = w_+(n)C(n) - w_-(n+1)C(n+1) = w_+(n)C^0(n)\left[\frac{C(n)}{C^0(n)} - \frac{C(n+1)}{C^0(n+1)}\right]. \tag{16.7}$$

For $n = 1$ there are plenty of isolated adatoms and one assumes that $C(1) = C^0(1)$. As for the boundary condition at $n = \infty$, the infinitely large nucleus disappears by the completion of a monatomic layer and thus $C(\infty) = 0$. Then by using the relation

$$\sum_{n=1}^{\infty} \frac{J}{w_+(n)C^0(n)} = \sum_{n=1}^{\infty}\left[\frac{C(n)}{C^0(n)} - \frac{C(n+1)}{C^0(n+1)}\right] = \frac{C(1)}{C^0(1)} - \frac{C(\infty)}{C^0(\infty)} = 1, \tag{16.8}$$

the nucleation flux J is obtained as

$$J = \left[\sum_{n=1}^{\infty} \frac{1}{w_+(n)C^0(n)}\right]^{-1}. \tag{16.9}$$

Since $C^0(n)^{-1} = c_\infty^{-1}\exp[G(n)/k_BT]$ has the maximum at the critical nucleus size n_c of Eq.(16.2) and $w_+(n)$ varies slowly in proportional to the surface area as $w_+(n) \sim n^{1/2}$, the dominant contribution to the summation in Eq.(16.9) comes near n_c. For small $\Delta\mu$, n_c is so large that the summation can be replaced by the integration as

$$\sum_n \frac{1}{w_+(n)C^0(n)} \approx \frac{1}{w_+(n_c)c_\infty}\int dn \exp\left(\frac{G_c + \frac{1}{2}G^{(2)}(n-n_c)^2}{k_BT}\right)$$

$$= \frac{1}{w_+(n_c)c_\infty}\exp\left(\frac{G_c}{k_BT}\right)\sqrt{\frac{2\pi k_BT}{|G^{(2)}|}}. \tag{16.10}$$

Here G_c is the maximum of the free energy cost given in Eq.(16.3) and $G^{(2)}$ is the second derivative of the activation free energy at n_c as

$$G^{(2)} = \left.\frac{d^2G}{dn^2}\right|_{n_c} = -\frac{1}{2}\beta\sqrt{\pi\Omega_2}n_c^{-3/2} = -\frac{\Delta\mu^3}{2\pi\Omega_2\beta^2}. \tag{16.11}$$

Therefore, the nucleation rate is written as

$$J \approx Zw_+(n_c)c_\infty\exp\left(-G_c/k_BT\right) \tag{16.12}$$

with the Zeldovich factor

$$Z = \sqrt{\frac{|G^{(2)}|}{2\pi k_B T}} = \left(\frac{\Delta\mu^3}{4\pi^2 \Omega_2 k_B T \beta^2}\right)^{1/2} = \left(\frac{\Delta\mu}{4\pi k_B T n_c}\right)^{1/2}. \tag{16.13}$$

Since the adatom capture by the nucleus takes place at the periphery of the nucleus, $w_+(n_c)$ is proportional to the perimeter length as [80, 69]

$$w_+(n_c) = \left.\frac{dn}{dt}\right|_{n_c} \approx v_{0,+} \frac{2\pi \rho_c}{\Omega_2} = 2x_s \Omega_2 f \frac{2\pi \beta}{\Delta\mu}, \tag{16.14}$$

where $v_{0,+} = 2x_s \Omega_2 f$ is the step advance rate by atom incorporation in Eq.(13.10). If there is a direct incorporation of atoms from the ambient gas phase, this contribution should be included in $v_{0,+}$ [69]. Thus

$$J = \frac{2x_s \Omega_2 f c_\infty}{a} \left(\frac{\Delta\mu}{k_B T}\right)^{1/2} \exp\left(-\frac{G_c}{k_B T}\right). \tag{16.15}$$

When the nucleation rate J is large, nucleation starts at various places on a crystal surface and those nuclei spread with the advance velocity v, coalesce and complete a single layer. This mode of layer growth is called the multinucleation growth. We now calculate the normal growth rate V of the crystal in this situation. On a flat surface it takes a time T for the completion of a single layer growth. During this time interval T, nucleation occurs with the rate J per unit area and time. At a time t after the nucleation started, its radius is vt and occupies the area $\pi(vt)^2$. Therefore, during T the area scanned by nuclei is calculated as

$$\int_0^T (SJ)\pi(vt)^2 dt = S\frac{\pi J}{3}v^2 T^3, \tag{16.16}$$

where S is the total area. When the scanned area (16.16) coincides with the total area S, the single layer growth is completed, and thus the time T is determined by

$$\frac{\pi J}{3}v^2 T^3 = 1. \tag{16.17}$$

After this time T the crystal surface moves up a distance a, and the normal growth rate by the two-dimensional nucleation mechanism is given by

$$V = \frac{a}{T} = a\left(\frac{\pi J v^2}{3}\right)^{1/3}. \tag{16.18}$$

Since the advance rate of the step is given in Eq.(13.10) or

$$v = 2x_s \Omega_2 f_{eq}\left[\exp\left(\frac{\Delta\mu}{k_B T}\right) - 1\right], \tag{16.19}$$

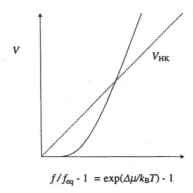

$$f/f_{\text{eq}} - 1 = \exp(\Delta\mu/k_BT) - 1$$

Figure 16.3: Schematic diagram of the growth velocity V by the two-dimensional nucleation.

the normal growth rate is

$$
\begin{aligned}
V &= 2x_{\text{s}}\Omega_2 f_{\text{eq}}\left(\frac{\pi c_{\text{eq}}\Omega_2}{3}\right)^{1/3}\exp\left(\frac{2\Delta\mu}{3k_BT}\right) \\
&\quad \times \left(\frac{\Delta\mu}{k_BT}\right)^{1/6}\left[\exp\left(\frac{\Delta\mu}{k_BT}\right) - 1\right]^{2/3}\exp\left(-\frac{\pi\beta^2\Omega_2}{3\Delta\mu k_BT}\right) \\
&\approx 2x_{\text{s}}\Omega_2 f_{\text{eq}}\left(\frac{\pi c_{\text{eq}}\Omega_2}{3}\right)^{1/3}\left(\frac{\Delta\mu}{k_BT}\right)^{5/6}\exp\left(-\frac{\pi\beta^2\Omega_2}{3\Delta\mu k_BT}\right).
\end{aligned}
\tag{16.20}
$$

The last approximation is obtained for small supersaturation $\Delta\mu$ where $c_\infty \approx c_{\text{eq}}$, and the result is schematically shown in Fig.16.3. Since V is proportional to the exponential of $\Delta\mu^{-1}$, it is very small for small $\Delta\mu$. The rate V becomes observable only when $G_{\text{c}}(\Delta\mu)$ is of the order or smaller than the thermal energy k_BT;

$$\Delta\mu > \Delta\mu_{\text{c}} \equiv \frac{\pi\beta^2\Omega_2}{3k_BT}. \tag{16.21}$$

For large enough $\Delta\mu$, the nucleation barrier vanishes, and the growth rate is $V \sim (\Delta\mu)^{1/6}\exp(4\Delta\mu/3k_BT)$, larger than the linear growth law $V \propto e^{\Delta\mu/k_BT} - 1$. This result is incorrect. For large $\Delta\mu$, the critical nuclear size n_{c} becomes so small that $n_{\text{c}} < 1$, and the crystal surface becomes kinetically rough. Then the growth rate should be that of the rough surface or the Hertz and Knudsen formula (3.4).

At the thermal roughening temperature T_R and above, the step free energy β vanishes, and the energy barrier G_{c} disappears. Thus the growth mode varies from the nucleation controlled exponential type, $V \sim \exp(-C/\Delta\mu)$ below the roughening transition temperature T_R, to the normal growth rate $V \sim \Delta\mu$ above T_R. This is used for the criterion to determine the roughening temperature in the Monte Carlo

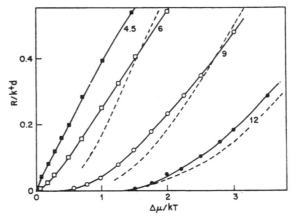

Figure 16.4: Growth rate R versus chemical potential difference of the SOS model. Numbers aside the curves represent inverse temperatures, $6J/k_BT$. k^+ here is the deposition rate, and d is the atomic height. The dashed lines are the growth rates calculated from the two-dimensional nucleation model [70]

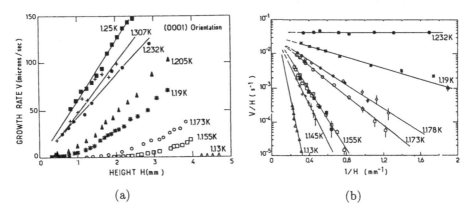

Figure 16.5: Growth rate of He crystal near and below the roughening temperature $T_R = 1.23K$ (a) in the normal and (b) in the semi-logarithmic plots [198]. H here is proportional to the chemical potential difference $\Delta\mu$.

simulation as is shown in Fig.16.4 [70], and in the experiment on He as shown in Fig.16.5 [198].

Exercise: When the nucleation rate J is very slow or the step advancement velocity v is very fast, there is only a single nucleus during the layer growth. Then the normal growth rate V differs from multinucleation case. Calculate the normal growth rate V of a surface with an area S. This growth mechanism is called the single nucleation growth.

Answer: When a single nucleus is created, it spreads over the whole surface S quickly before further nuclei are formed on the same level. The time T necessary for the completion of the normal growth of a height a satisfies the relation

$$JST = 1. \tag{16.22}$$

The growth rate is obtained as

$$V = \frac{a}{T} = aSJ = 2x_s\Omega_2 fSc_\infty \left(\frac{\Delta\mu}{k_BT}\right)^{1/2} \exp\left[-\frac{\pi\beta^2\Omega_2}{\Delta\mu k_BT}\right]. \tag{16.23}$$

If one plots $\ln V$ versus $1/\Delta\mu$, the slope by a single nucleation mechanism is three times larger than that by a multiple nucleation mechanism. This crossover is said to be observed in polymer crystallization [81, 179].

17   Asymmetry in Attachment Kinetics

In the treatment of adatom diffusion, Burton, Cabrera and Frank (BCF) [43] assumed the fast kinetics at the step such that the local equilibrium is realized there. Also they assumed the symmetry in the attachment kinetics of adatoms from the upper and lower terraces at the step. However, the incorporation rate of an adatom from an upper terrace needs not to be same with that from a lower terrace as shown in Fig.17.1 [172, 190]: From the upper terrace an adatom has to break many chemical bonds with the underlying substrate atoms when it crosses over the step down to the crystallization position. From the lower terrace, on the other hand, an adatom

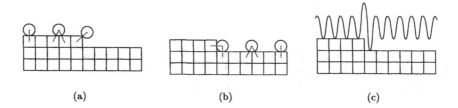

Figure 17.1: Asymmetry in the incorporation of an adatom (a) from the upper and (b) from the lower terraces, called Schwoebel effect. (c) The potential energy profile of the adsorbed atom.

can simply make additional bonds with the step atoms before it reaches kink sites. Additional energy barrier for the crystallization makes the kinetic coefficient from the upper terrace K_- smaller than that from the lower terrace K_+. This asymmetry in attachment kinetics at the step is first studied by Schwoebel and Shipsey and is called the Schwoebel effect [172].

17.1   Schwoebel Effect

We consider the step down configuration, where the step is running on average in x direction at $y = 0$ and the terrace in front at $y > 0$ is lower than the terrace in the back $y < 0$. Since the step is thermally rough, the step advance rates, v_+ and v_-, by the adatom incorporation from the upper and the lower terraces respectively are linearly proportional to the supersaturation as

$$v_\pm = K_\pm \left(c_\pm - c_{eq} - c_{eq} \frac{\tilde{\beta}\Omega_2}{k_B T} \kappa \right). \tag{17.1}$$

Here the Gibbs-Thomson effect of curvature is included. The Schwoebel effect means that the kinetic coefficients are different: $K_+ \neq K_-$.

We now restrict ourselves in the extreme and the simple case that

$$K_- = 0 \qquad \text{and} \qquad K_+ = \infty. \tag{17.2}$$

In this case, there is no crystallization from the upper back terrace. Since the crystallization takes place by the atom incorporation only from the lower terrace, it is called a one-sided model. Furthermore, the kinetics from the lower front terrace is assumed extremely fast such that the local equilibrium is realized: The adatom density in front of the step c_+ is equal to the equilibrium value c_{eq} with the Gibbs-Thomson effect

$$c_+ = c_{eq} \left(1 + \frac{\tilde{\beta}\Omega_2}{k_B T} \kappa \right). \tag{17.3}$$

As in the BCF model, the atoms are deposited on the crystal surface with a flux f, the adsorbed atoms then diffuse on the surface with a surface diffusion constant D_s, and then evaporates back into an ambient vapor after a life time τ. For a straight step its advance velocity v_0 is calculated to be

$$v_0 = (f - f_{eq}) x_s \Omega_2, \tag{17.4}$$

where $f_{eq} = c_{eq}/\tau$ is the equilibrium deposition flux, $x_s = \sqrt{D_s \tau}$ is the surface diffusion length and Ω_2 the atomic area. The velocity v_0 is half of that given in (13.10) in BCF theory, since there is a contribution only from the front terrace. Eq.(17.4) shows that the step incorporates atoms deposited in the range x_s in front of the step.

We now consider the stability of the straight step [8]. If the step is pushed forward at some part by fluctuation, the region to incorporate adatoms expands radially as

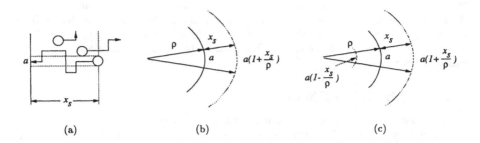

Figure 17.2: Capturing region of adsorbed atoms of a one-sided model in front of a step (a) for a straight step and (b) for a curved step. (c) Capturing region for a symmetric model for a curved step.

is shown in Fig.17.2. The area of the capturing region increases approximately by a factor $1 + x_s/2\rho$, and the velocity increases with the same factor. Thus the step with a curvature $\kappa = 1/\rho$ has a velocity higher than the straight one by

$$\delta v_d = v_0 \cdot \frac{x_s}{2}\kappa. \tag{17.5}$$

The bump is accelerate compared to the straight part and is pushed further forward. The diffusion causes a destabilization of a step profile.

Competing with this destabilization is the stabilization effect by the step stiffness. Since the equilibrium density increases at a curved part, the deposition flux to maintain equilibrium at a curved step increases as $f_{eq}(1 + \tilde{\beta}\Omega_2\kappa/k_BT)$, and the driving force or the supersaturation decreases correspondingly. The velocity of a bump decreases by

$$\delta v_s = -f_{eq}\frac{\tilde{\beta}\Omega_2}{k_BT}\kappa x_s\Omega_2. \tag{17.6}$$

Both destabilizing and stabilizing effects are, in the first order approximation, proportional to the curvature, as is apparent in Eq.(17.5) and (17.6). Since the diffusional instability increases with the velocity v_0, the instability wins eventually by increasing the deposition rate. The instability takes place when $\delta v_d + \delta v_s = 0$. The critical deposition rate f_c is determined from the Eqs. (17.4-17.6) as

$$f_c = f_{eq}\left(1 + \frac{2\tilde{\beta}\Omega_2}{x_s k_BT}\right). \tag{17.7}$$

Exercise: Explain that in the BCF model where $K_+ = K_- = \infty$, the step instability does not take place.

Answer: Consider a part of a step pushed forward with a positive radius ρ as shown in Fig.17.2c. The velocity contribution from the lower terrace increases as $v_+ = v_0(1 + x_s/2\rho)$, but that from the upper terrace decreases as $v_- = v_0(1 - x_s/2\rho)$. The total velocity $v_+ + v_- = 2v_0$ remains independent of the step deformation in the stationary approximation.

17.2 Structure of an unstable step

In order to describe the profile of a destabilized step, one has to treat the problem quantitatively and analytically. First we treat the stability of a straight step in a linear analysis [8, 186]. The step deformation $y = \zeta(x, t)$ is decomposed in Fourier modes. In the linear analysis there is no coupling among modes, and the consideration of a single mode with a wavenumber q is sufficient:

$$\zeta(x, t) = v_0 t + a_q e^{\omega_q t} \cos qx. \tag{17.8}$$

Here v_0 is the velocity of a straight step moving in y direction. If the amplification rate ω_q of the deformation is negative, the amplitude of the deformation diminishes and the straight step recovers. On the contrary, if ω_q is positive, the mode amplitude increases and the straight step is unstable.

The normal direction of this deformed step (17.8) is given by

$$\mathbf{n} = \frac{(-\partial\zeta/\partial x, 1)}{\sqrt{1 + (\partial\zeta/\partial x)^2}} \approx (qa_q e^{\omega_q t} \sin qx, 1), \tag{17.9}$$

the curvature is

$$\kappa = \frac{-\partial^2\zeta/\partial x^2}{[1 + (\partial\zeta/\partial x)^2]^{3/2}} \approx q^2 a_q e^{\omega_q t} \cos qx, \tag{17.10}$$

and the growth velocity is obtained as

$$\mathbf{v} = (0, \partial\zeta/\partial t) = (0, v_0 + \omega_q a_q e^{\omega_q t} \cos qx). \tag{17.11}$$

Therefore the normal velocity is written up to the first order of a_q as

$$v_n = (\mathbf{n} \cdot \mathbf{v}) = v_0 + \omega_q a_q e^{\omega_q t} \cos qx. \tag{17.12}$$

The adatom density $c(x, y; t)$ is also modified with a wavenumber q in x direction. For $y \to \infty$, the modification should decay and thus the density is written as

$$c(x, y; t) = c_\infty + (c_{eq} - c_\infty)e^{-(y-v_0 t)/x_s} + \delta c_q e^{\omega_q t} \cos qx\, e^{-\Lambda_q(y-v_0 t)} \tag{17.13}$$

in front of the step. The first two terms are the density distribution around the straight step. Since the density satisfies the diffusion equation in the stationary approximation,

$$\nabla^2 c + \frac{c_\infty - c}{x_s^2} = 0, \tag{17.14}$$

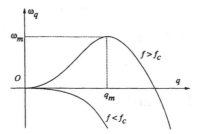

Figure 17.3: Dispersion relation of a step deformation with Schwoebel effect.

the damping coefficient Λ_q in y direction satisfies the relation $-q^2 + \Lambda_q^2 - x_s^{-2} = 0$ or

$$\Lambda_q = \sqrt{q^2 + x_s^{-2}}. \tag{17.15}$$

From the local equilibrium condition at the step, Eq.(17.3), the density deviation δc_q is obtained as

$$\delta c_q = -a_q \left(\frac{c_\infty - c_{eq}}{x_s} - \frac{c_{eq}\Omega_2 \tilde{\beta}}{k_B T} q^2 \right), \tag{17.16}$$

where the curvature κ is approximated in the linear approximation (17.10). From the continuity boundary condition

$$\Omega_2^{-1} v_n = D_s \frac{\partial c}{\partial n} = D_s \left(\frac{\partial c}{\partial x} n_x + \frac{\partial c}{\partial y} n_y \right), \tag{17.17}$$

one obtains the relation

$$\Omega_2^{-1} \left(v_0 + \omega_q a_q e^{\omega_q t} \cos qx \right) \approx D_s \left[\frac{c_{eq} - c_\infty}{-x_s} e^{-(\zeta - v_0 t)/x_s} - \Lambda_q \delta c_q e^{\omega_q t} \cos qx \, e^{-\Lambda_q(\zeta - v_0 t)} \right]$$

$$\approx \frac{D_s}{x_s}(c_\infty - c_{eq}) \left(1 - \frac{a_q}{x_s} e^{\omega_q t} \cos qx \right) - D_s \Lambda_q \delta c_q e^{\omega_q t} \cos qx. \tag{17.18}$$

By comparing the zeroth order term in a_q, one obtains the velocity v_0 of the straight step (17.4). From the first order of a_q, the dispersion relation between ω_q and q is obtained as

$$\omega_q = -D_s \Omega_2 \left[\frac{c_\infty - c_{eq}}{x_s^2} - \Lambda_q \left(\frac{c_\infty - c_{eq}}{x_s} - \frac{c_{eq}\Omega_2 \tilde{\beta}}{k_B T} q^2 \right) \right]$$

$$= v_0(\Lambda_q - x_s^{-1}) - D_s \Omega_2^2 \Lambda_q \frac{c_{eq}\tilde{\beta}}{k_B T} q^2, \tag{17.19}$$

which is shown in Fig.17.3. The first term corresponds the diffusional destabilization or δv_d in (17.15), and the second term corresponds to the energetic stabilization or

δv_{s} in (17.6) for long wavelength or small q ($\ll x_{\mathrm{s}}$). The dispersion relation (17.19) can be further rearranged as

$$\omega_q = -\mu_q \cdot \nu_q \tag{17.20}$$

with the mobility

$$\mu_q = D_{\mathrm{s}} \Omega_2^2 \Lambda_q \frac{c_{\mathrm{eq}}}{k_B T} (> 0) \tag{17.21}$$

and the restoring force

$$\begin{aligned}
\nu_q &= \tilde{\beta} q^2 - \frac{k_B T}{\Omega_2 x_{\mathrm{s}}} \left(\frac{f}{f_{\mathrm{eq}}} - 1 \right) \left(1 - \frac{1}{\Lambda_q x_{\mathrm{s}}} \right) \\
&\approx \tilde{\beta}_{\mathrm{eff}} q^2 + \frac{k_B T x_{\mathrm{s}}^3}{4 \Omega_2} \left(\frac{f}{f_{\mathrm{eq}}} - 1 \right) q^4 + \cdots,
\end{aligned} \tag{17.22}$$

where the long wavelength limit, $q x_{\mathrm{s}} \ll 1$, is used in the last approximation. The effective stiffness $\tilde{\beta}_{\mathrm{eff}}$ is the force constant for the step recovery modified by the diffusion destabilizing effect as

$$\tilde{\beta}_{\mathrm{eff}} = \tilde{\beta} \left[1 - \frac{x_{\mathrm{s}} k_B T}{2 \tilde{\beta} \Omega_2} \left(\frac{f}{f_{\mathrm{eq}}} - 1 \right) \right] = \tilde{\beta} \frac{f_{\mathrm{c}} - f}{f_{\mathrm{c}} - f_{\mathrm{eq}}}. \tag{17.23}$$

Here f_{c} is the critical deposition rate defined in the previous subsection, (17.7). At the instability point f_{c}, the effective stiffness vanishes, and the step looses the restoring force to the straight form. Equation (17.23) shows that the step stiffness depends on the deposition rate f. For small f $\tilde{\beta}_{\mathrm{eff}}$ is large and the step is stiff, while for large f it is soft. This variation of the step stiffness is observed in the experiment on Si [125, 126].

Near the instability point f_{c}, we have a small parameter

$$\epsilon = \frac{f_{\mathrm{c}} - f}{f_{\mathrm{c}} - f_{\mathrm{eq}}}. \tag{17.24}$$

For positive ϵ, the dispersion relation for long wavelength has a positive maximum at q_{m} of order $\epsilon^{1/2}$ as shown in Fig.17.3. The maximum value of ω_q is of order ϵ^2. This mean that near the instability point, the step modulation has a large spatial extension of order $\epsilon^{-1/2}$, and relaxes slowly in a time of order ϵ^{-2}. In the weakly nonlinear case the dynamics of an unstable step can be analyzed by rescaling the variables and extract slow dynamics [199, 13]. One scales space and time as follows;

$$\begin{aligned}
X &= \epsilon^{1/2} x / x_{\mathrm{s}} \\
Y &= y / x_{\mathrm{s}} \\
T &= \epsilon^2 t / \tau,
\end{aligned} \tag{17.25}$$

and introduce a dimensionless field u and an interface position ζ, which are expanded in terms of ϵ as

$$\begin{aligned}
u &= \Omega_2 (c - c_\infty) = u_0 + \epsilon u_1 + \epsilon^2 u_2 + \cdots \\
\frac{\zeta(x)}{x_{\mathrm{s}}} &= \epsilon H = \epsilon H_0 + \epsilon^2 H_1 + \cdots
\end{aligned} \tag{17.26}$$

The diffusion equation is described in these variables as

$$\epsilon u_{XX} + u_{YY} - u = 0. \tag{17.27}$$

Here and later in this section, the subscript X or Y means the derivative by them. The boundary conditions are written as

$$V_c(1 + \epsilon) + \epsilon^3 H_T = u_Y - \epsilon^2 u_X H_X \tag{17.28}$$

$$u_s = -V_c(1 + \epsilon) - \frac{V_c}{2} \frac{\epsilon^2 H_{XX}}{[1 + \epsilon^3 (H_X)^2]^{3/2}}, \tag{17.29}$$

where $V_c = 2\Omega_2^2 c_{eq} \tilde{\beta}/k_B T x_s$ is the step velocity at the critical point f_c. By comparing each order of ϵ, one obtains the solutions and relations as

$$O(\epsilon^0): \quad u_0 = A_0 e^{-Y}, A_0 = -V_c \tag{17.30}$$

$$O(\epsilon^1): \quad u_1 = A_1 e^{-Y}, \frac{A_1}{A_0} = 1 + H_0(X,T) \tag{17.31}$$

$$O(\epsilon^2): \quad u_2 = A_2 e^{-Y} + \frac{1}{2} A_{1,XX} Y e^{-Y}, \frac{A_2}{A_0} = H_0 + \frac{1}{2} H_0^2 + \frac{1}{2} H_{0,XX} + H_1 \tag{17.32}$$

$$O(\epsilon^3): \quad u_3 = A_3 e^{-Y} + \frac{1}{2} A_{2,XX} Y e^{-Y} + \frac{1}{8} A_{1,XXXX}(Y^2 + Y) e^{-Y},$$

$$\frac{A_3}{A_0} = \frac{1}{2} H_0^2 + \frac{1}{6} H_0^3 + H_1 + H_0 H_1 + \frac{1}{2} H_{1,XX} + H_2. \tag{17.33}$$

To fulfill the boundary conditions (17.28) and (17.29) consistently at the order $O(\epsilon^3)$, H_0 should satisfy the differential equation as

$$-\frac{1}{V_c} \frac{\partial H_0}{\partial T} = \frac{1}{2} \frac{\partial^2 H_0}{\partial X^2} + \frac{3}{8} \frac{\partial^4 H_0}{\partial X^4} - \frac{1}{2} \left(\frac{\partial H_0}{\partial X} \right)^2. \tag{17.34}$$

The linear part represents the dispersion relation (17.20-17.22) in this unit. The nonlinear term $(\partial H_0/\partial X)^2$ is the lowest possible term compatible with the translational symmetry of the system. There cannot be terms containing H_0 such as H_0^2 or $H_0 \partial H_0/\partial X$ since the evolution should be independent of the absolute position of the step itself. The equation (17.34) is called the Kuramoto-Sivashinsky equation [118, 177] and is known to have spatio-temporal chaos, as shown in Fig.17.4. The front has many hills and valleys. Valleys shift randomly to the left or to the right, collide with each other and annihilate. A hill widely spread splits and a new valley is formed randomly. The same chaotic behavior is observed in the Monte Carlo simulation of the step advancing in the adatom diffusion field, as shown in Fig.17.5 [170].

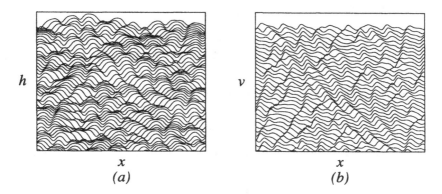

Figure 17.4: Time evolution of the Kuramoto-Sivashinsky equation, showing spatio-temporal chaotic behavior. (a) The height h and (b) the gradient $v = \partial h / \partial x$.

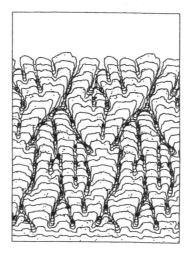

Figure 17.5: Monte Carlo simulation of the time evolution of an unstable step growing from the diffusing adsorbed atoms [170]. It shows the spatio-temporal chaos similar to Fig.17.4.

Part IV

Diffusion-Limited Growth: Pattern Formation

So far we considered the effect of surface structure and surface kinetics on the crystal growth. But there are other processes which controls the growth. Crystallization proceeds in the following sequences:

1. Atoms to be crystallized are transported to the crystal surface (= chemical diffusion),

2. they are incorporated in the crystal at the surface (=surface kinetics),

3. and the released latent heat should be transported away from the crystal surface (= heat conduction).

If all these processes are fast enough, ideal growth laws can be realized. But in reality, the slowest process governs the growth rate as a whole, and the deviation from the ideal linear growth laws explained in part I is expected. For the crystal with an atomically flat interface, the second, surface kinetic process is the slowest one and it controls the growth. The growth law is found to differ from the ideal one, as explained in Part III. For rough surfaces, kinetics is fast, but still the growth is different from the ideal behavior when the transport processes (1) and (3) are slow. In this part we consider the crystal growth and its morphology, when the growth is controlled by chemical or heat diffusion processes.

Diffusion field induces instability in the growing interface, as, for example, shown in the previous section 17. This diffusional instability causes variety of patterns in the growth shape of the crystal. Examples are the fractal structure of the diffusion-limited aggregation (Fig.1.3), dendrite in the melt or solution growth (Fig.1.2), lamellar structure in the eutectic growth (Fig.26.2). These topics are studied in the present part.

18   Diffusion Equation

Since the mass or energy is conserved, the material or heat transport follows a diffusion equation [122]. To consider the problem concretely, we consider the crystal growth from the undercooled melt where the heat conduction evacuates the produced latent heat from the crystal surface. Since the melt is cooled at a temperature T_∞ below the melting temperature T_M, the Gibbs free energy of the crystal per volume G_S is lower than that of the melt G_L as (3.1) or

$$\Delta G = G_L - G_S = L \frac{T_M - T_\infty}{T_M}, \tag{18.1}$$

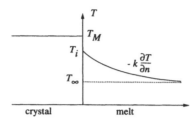

Figure 18.1: Temperature profile around the flat interface of the crystal growing in the supercooled melt.

where L is the latent heat per volume. This ΔG drives the material to crystallize. By the crystallization, however, the latent heat L released at the interface heats up the crystal surface to a temperature T_i, higher than the far field value T_∞. Therefore, the driving force of crystallization at the interface reduces to $\Delta G_i = L(T_M - T_i)/T_M$. For many metals or some plastic materials, the interface is rough at the melting point and thus the growth rate is proportional to ΔG_i, following the Wilson-Frenkel formula (3.5) or

$$V_n = K_T(T_M - T_i) \tag{18.2}$$

with the kinetic coefficient K_T. The remaining supercooling $T_i - T_\infty$ drives the heat transport and prohibits the interface from being heated up by the latent heat. The heat transport in liquid is described by the heat conduction equation

$$C_p \frac{\partial T}{\partial t} = k\nabla^2 T, \tag{18.3}$$

where C_p is the specific heat per volume, k is the thermal conductivity. When the crystal grows with a normal velocity V_n the latent heat LV_n is produced per unit area in unit time. This heat should be transported away by the heat flow in the normal direction $\mathbf{n}$ as

$$LV_n = -k(\mathbf{n} \cdot \nabla)T \equiv -k\partial_n T. \tag{18.4}$$

Here we restrict ourselves to the one-sided model such that the heat is transported only in the liquid. Extension to the two-sided or symmetric models is straightforward. The driving force $\Delta G = L(T_M - T_\infty)/T_M$ is partitioned into the kinetic part ΔG_i and the transport part $\Delta G_t = L(T_i - T_\infty)/T_M$. The interface temperature T_i is so determined that the velocity determined from the kinetics (18.2) agrees with that from the transport (18.4).

The fundamental equations (18.2-18.4) is derived for the melt growth where the heat conduction controls the growth. Similarly, in the solution chemical diffusion controls the growth of crystal. In order to generalize the situation and to take up the essential features of the problem, we scale variables in dimensionless form. The dimensionless diffusion field is defined as

$$u(\mathbf{x}, t) = \frac{T(\mathbf{x}, t) - T_\infty}{L/C_p}, \tag{18.5}$$

and it follows the diffusion equation

$$\frac{\partial u}{\partial t} = D\nabla^2 u \tag{18.6}$$

with the thermal diffusivity $D = k/C_p$. Boundary conditions are the conservation law

$$V_n = -D\partial_n u, \tag{18.7}$$

and the Wilson-Frenkel law at the interface

$$V_n = K'(\Delta - u_i - d\kappa), \tag{18.8}$$

where $K' = K_T(L^2/C_p T_\mathrm{M})$ is the new kinetic coefficient, which will be denoted K hereafter.

$$\Delta = \frac{T_\mathrm{M} - T_\infty}{L/C_p} \tag{18.9}$$

is the dimensionless undercooling, which is the undercooling normalized by the temperature increase caused by the latent heat production. $u_i = (T_i - T_\infty)/LC_p^{-1}$ is the value of the diffusion field at the interface,

$$d = \frac{\tilde{\gamma}}{L}\frac{T_\mathrm{M}}{L/C_p} \tag{18.10}$$

is the capillary length proportional to the surface stiffness $\tilde{\gamma}$, and κ is the curvature.

For the rough interface with many steps and kinks, the kinetics is expected to be very fast. In the limit of $K \to \infty$, the local equilibrium is realized at the interface as

$$u_i = \Delta - d\kappa, \tag{18.11}$$

instead of the Wilson-Frenkel law (18.8). In this case, the crystal growth is totally governed by the diffusion.

19 Flat Interface

When the crystal grows steadily with a macroscopically flat interface in z direction, the growth velocity V is expected to depend on the undercooling Δ. In the coordinate frame comoving with the crystal $(x, y, z' = z - Vt)$, the diffusion equation (18.6) is transformed as

$$\frac{1}{D}\frac{\partial u}{\partial t} = \nabla^2 u + \frac{2}{l_D}\frac{\partial u}{\partial z'}, \tag{19.1}$$

where

$$l_D = \frac{2D}{V} \qquad (19.2)$$

is the diffusion length. For a steadily growing crystal, the diffusion field does not vary in time in the moving frame, $\partial u/\partial t = 0$, or

$$\nabla^2 u + \frac{2}{l_D} \frac{\partial u}{\partial z'} = 0. \qquad (19.3)$$

Taking the origin of z' coordinate on the flat interface, the diffusion field is solved as

$$u(z') = A \exp\left(-\frac{2z'}{l_D}\right) \qquad (19.4)$$

under the far field condition $u(z' = \infty) = 0$. The diffusion length l_D is the thickness of the diffusion layer, and characterizes the spatial variation of the diffusion field. From the conservation boundary condition (18.7) one obtains the relation

$$V = -D\partial_{z'} u = D\frac{2A}{l_D} = AV. \qquad (19.5)$$

Thus, $A = 1$. From the kinetic boundary condition (18.8), the growth velocity is determined as

$$V = K(\Delta - 1). \qquad (19.6)$$

Since $u_i = A = 1$, the interface temperature in the conventional unit is $T_i = T_\infty + L/C_p$ by utilizing Eq.(18.5): The interface is heated up by the latent heat. Since T_i should be colder than the melting temperature T_M for the crystal to grow, Δ has to be larger than 1. Even though the melt is supercooled at $\Delta > 0$, the crystal with a flat interface cannot grow steadily for $\Delta < 1$. The latent heat released heats up the interface so high that the heat conduction cannot transport it quick enough with small supercooling. In the case of local equilibrium ($K = \infty$), the time-dependent solution of the one dimensional problem is exactly obtained, and the growth velocity decreases as $t^{-1/2}$.

Exercise: We consider the crystal growth with a flat interface under the local equilibrium condition. When the interface is moving in z direction as $z = \zeta(t)$, the growth velocity is shown [99] to decreases as $t^{-1/2}$ by assuming the scaling form for the diffusion field as

$$u(z,t) = u\left(\frac{z}{\zeta(t)}\right). \qquad (19.7)$$

(1) Show that the time-dependent diffusion equation (18.6) is transformed to the ordinary differential equation for $u(w)$ with a new variable $w = z/\zeta(t)$ as

$$\frac{d^2 u}{dw^2} + \frac{\zeta \dot{\zeta}}{D} w \frac{du}{dw} = 0. \qquad (19.8)$$

(2) If u depends only on w, then show that

$$\zeta(t) = \sqrt{4DPt} \tag{19.9a}$$

$$u(w) = 2Pe^P \int_w^\infty e^{-Pw^2} dw, \tag{19.9b}$$

where P is defined by the relation

$$\Delta = \sqrt{\pi P} e^P \text{erfc}(\sqrt{P}), \tag{19.10}$$

with the error function defined by

$$\text{erfc}(\sqrt{P}) = \frac{2}{\sqrt{\pi}} \int_{\sqrt{P}}^\infty e^{-x^2} dx. \tag{19.11}$$

Eq.(19.9a) shows that the interface velocity decreases to zero as $\dot{\zeta} \propto t^{-1/2}$.

Answer:

(1) By changing variables, derivatives are related as

$$\frac{\partial u(z/\zeta(t))}{\partial t} = \frac{\partial w}{\partial t} \frac{du}{dw} = -\frac{z\dot{\zeta}}{\zeta^2} \frac{du}{dw} \tag{19.12}$$

$$\frac{\partial u(z/\zeta(t))}{\partial z} = \frac{\partial w}{\partial z} \frac{du}{dw} = \frac{1}{\zeta} \frac{du}{dw} \tag{19.13}$$

$$\frac{\partial^2 u(z/\zeta(t))}{\partial z^2} = \frac{1}{\zeta^2} \frac{d^2u}{dw^2}, \tag{19.14}$$

where $\dot{\zeta} = d\zeta/dt$. Inserting the relations (19.12-19.14) in the diffusion equation (18.6), one easily ends with (19.8).

(2) By rearrangement, Eq.(19.8) can be written as

$$\frac{\zeta\dot{\zeta}}{D} = -\frac{d^2u/dw^2}{wdu/dw} = 2P. \tag{19.15}$$

The first term depends only on time t and the middle term only on w, and thus should both be constant, which is set $2P$. From the first equation, $\zeta\dot{\zeta} = 2DP$, the interface advances in proportional to $t^{1/2}$ as in (19.9). Integrating the second equality, one gets

$$\ln \frac{du}{dw} = -Pw^2 + const \tag{19.16}$$

or

$$\frac{du}{dw} = Ce^{-Pw^2}. \tag{19.17}$$

With the far field condition, $u(\infty) = 0$, it is integrated as

$$u(w) = -C \int_w^\infty e^{-Pw^2} dw. \tag{19.18}$$

From the local equilibrium condition (18.11) at the interface, $z = \zeta(t)$ or $w = 1$, the integral constant C is determined as

$$\Delta = -C \int_1^\infty e^{-Pw^2} dw = -\frac{C}{2}\sqrt{\frac{\pi}{P}}\mathrm{erfc}(\sqrt{P}). \tag{19.19}$$

The conservation law (18.7) is now written as

$$\dot{\zeta} = -DC \left(e^{-Pw^2} \frac{\partial w}{\partial z} \right)_{w=1} = -DC\frac{e^{-P}}{\zeta}. \tag{19.20}$$

By inserting (19.15) and (19.19) in (19.20) one gets (19.10).

20   Spherical Crystal

When the crystal grows in a spherical shape, it emits heat in all directions, and it may sustain steady growth. But it will be shown that the steady growth is not possible in this case either. We study the time evolution of the radius R and its radial velocity $\dot{R} = dR/dt$, as shown in Fig.20.1. From the symmetry the diffusion field u is expected to depend only on the radial variable r in a spherical coordinate. Then the diffusion equation is written as

$$\frac{1}{D}\frac{\partial u}{\partial t} = \left(\frac{\partial^2}{\partial r^2} + \frac{2}{r}\frac{\partial}{\partial r} \right) u \approx 0. \tag{20.1}$$

In the last equality the stationary approximation is used, where the relaxation of diffusion field is very quick compared to the shape variation of the crystal. The stationary distribution of the diffusion field takes the form

$$u(r) = A/r \tag{20.2}$$

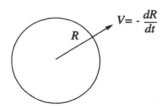

Figure 20.1: Crystal growth in spherical shape.

with an integration constant A. The far field condition $u = 0$ at $r \to \infty$ is satisfied in (20.2). The growth velocity V is determined by two ways, from Eq.(18.7) and (18.8), as

$$V = \frac{DA}{R^2} = K\left(\Delta - \frac{A}{R} - \frac{2d}{R}\right).$$
(20.3)

Here the relation of the curvature $\kappa = 2/R$ with the radius of the sphere R is used. The integration constant A is then determined as

$$A = \frac{R^2(\Delta - 2d/R)}{R + D/K},$$
(20.4)

and the growth velocity as

$$\frac{dR}{dt} = V = \frac{D(\Delta - 2d/R)}{R + D/K} = \frac{\Delta}{K^{-1} + R/D}\left(1 - \frac{\rho_c}{R}\right).$$
(20.5)

If the crystal radius R is smaller than the critical radius $\rho_c = 2d/\Delta$, sphere cannot grow. If R is greater than ρ_c, the sphere can grow. For small radius, $\rho_c \ll R \ll D/K$, the kinetics resistivity K^{-1} against the growth is dominant, and the growth velocity $K\Delta(1 - \rho_c/R)$ corresponds to that obtained phenomenologically in Eq.(7.2). When the sphere grows large with $R \gg D/K$, the diffusional resistivity R/D becomes dominant, and the growth rate varies as

$$V \approx \frac{D}{R}\Delta\left(1 - \frac{\rho_c}{R} - \frac{1}{K}\frac{D}{R} + \cdots\right) \approx \frac{D}{R}\Delta\left[1 - \frac{1}{R}\left(\rho_c + \frac{D}{K}\right)\right].$$
(20.6)

For sufficiently large sphere as $R \gg \rho_c$, the growth rate $V = \dot{R}$ is inversely proportional to the radius R, and by integration one gets

$$R^2(t) = \Delta D t + R^2(0).$$
(20.7)

Asymptotically as $t \to \infty$, the radius increases in proportional to $t^{1/2}$, and the velocity decreases as $V = \dot{R} \propto t^{-1/2}$. Thus the steady growth of sphere is impossible.

21   Parabolic Crystal

In the analysis so far, neither the planar nor the spherical crystals can grow steadily if the diffusion controls the growth. Is there any shape which allows the steady growth of a crystal? Ivantsov [89] has shown exactly that the needle crystal with a parabolic tip can grow steadily, if the crystal interface is in local equilibrium with infinitely fast kinetics and the surface tension is neglected, namely the crystal interface is the isotherm at the melting temperature T_M.

If the parabolic crystal grows steadily as

$$z = Vt - \frac{x^2 + y^2}{2R} = Vt - \frac{r^2}{2R}$$
(21.1)

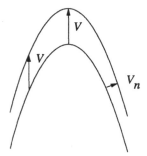

Figure 21.1: Crystal growth in parabolic needle shape. The latent heat is emitted dominantly near the tip.

as shown in Fig.21.1, its height increases in proportional to the elapsed time Δt, whereas the width increases only in proportional to $\sqrt{\Delta t}$. Therefore, the growth takes place mainly at the tip region, and the latent heat is emitted at the tip. The total amount of the latent heat released from the tip of parabolic dendrite down to the tail with a radius r is $L\pi r^2 V$ per time. The height of the dendrite is $r^2/2R$ and the surface area is proportional to r^3/R, precisely calculated to be $\pi r^3/3R$. Therefore the average heating per area is $3LVR/r$, which decreases for a long parabola at $r \to \infty$. Heat will not accumulate and a steady growth is allowed in this shape.

We now treat the problem more quantitatively. When the crystal is growing steadily in z direction with a velocity V, one transforms as usual to the moving frame of reference as $(x, y, z' = z - Vt)$. For the present shape it is convenient to transform further to the parabolic coordinate (Fig.21.2)

$$\begin{aligned}
\xi &= r - z', \\
\eta &= r + z', \\
\theta &= \arctan(y/x)
\end{aligned} \tag{21.2}$$

with $r = \sqrt{x^2 + y^2 + z'^2}$. (See Appendix A21). Then the diffusion equation in the steady state is described as

$$\frac{1}{\eta + \xi}\left(\frac{\partial}{\partial\eta}\eta\frac{\partial u}{\partial\eta} + \frac{\partial}{\partial\xi}\xi\frac{\partial u}{\partial\xi}\right) + \frac{1}{4\eta\xi}\frac{\partial^2 u}{\partial\theta^2} + \frac{1}{l_D}\frac{1}{\eta + \xi}\left(\eta\frac{\partial u}{\partial\eta} - \xi\frac{\partial u}{\partial\xi}\right) = 0, \tag{21.3}$$

where $l_D = 2D/V$ is the diffusion length defined in (19.2).

By denoting the interface as $\eta = \eta_i(\xi, \theta, t)$, the conservation boundary condition (18.7) is written as

$$\eta_i + \xi\frac{\partial\eta_i}{\partial\xi} + \frac{\eta_i + \xi}{2V}\frac{\partial\eta_i}{\partial t} = -l_D\left(\eta_i\frac{\partial u}{\partial\eta} - \xi\frac{\partial\eta_i}{\partial\xi}\frac{\partial u}{\partial\xi} - \frac{\eta_i + \xi}{4\eta_i\xi}\frac{\partial\eta_i}{\partial\theta}\frac{\partial u}{\partial\theta}\right), \tag{21.4}$$

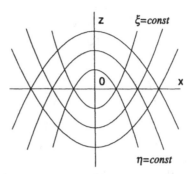

Figure 21.2: Parabolic coordinate.

and the local equilibrium condition (18.11) is written as

$$u(\eta_i) = \Delta, \tag{21.5}$$

since there is no capillary effect ($d = 0$).

The paraboloid crystal of revolution (21.1) corresponds to the interfacial shape with a constant η as $\eta_i = R$. Due to the symmetry, the field u depends only on η, and the diffusion equation reduces to

$$\frac{\partial}{\partial \eta}\left(\eta \frac{\partial u}{\partial \eta}\right) + \frac{1}{l_D}\left(\eta \frac{\partial u}{\partial \eta}\right) = 0. \tag{21.6}$$

It is easily solved as

$$u(\eta) = \Delta + C \int_R^\eta \eta^{-1} e^{-\eta/l_D} d\eta. \tag{21.7}$$

Here the local equilibrium boundary condition (21.5) is used. From the far field condition that $u(\eta \to \infty) = 0$, the integration constant C is determined as

$$C = -\frac{\Delta}{\int_R^\infty \eta^{-1} e^{-\eta/l_D} d\eta}. \tag{21.8}$$

From the continuity boundary condition (21.4) with $\eta_i = R$ and $\partial\eta_i/\partial\xi = \partial\eta_i/\partial\theta = \partial\eta_i/\partial t = 0$, one gets the relation

$$R = -l_D R \frac{\partial u}{\partial \eta}. \tag{21.9}$$

Inserting Eq.(21.7) and (21.8) into the relation (21.9), one gets the Ivantsov relation

$$\Delta = (R/l_D)e^{R/l_D} \int_R^\infty \eta^{-1} e^{-\eta/l_D} d\eta = Pe^P E_1(P) \tag{21.10}$$

with the Peclet number P defined as

$$P = \frac{R}{l_D} = \frac{RV}{2D}.$$ (21.11)

Here $E_1(P) = \int_P^\infty x^{-1} e^{-x} dx$ is the exponential integral function [1]. In two dimensions, one obtains the steadily growing parabolic crystal

$$z - vt = -\frac{x^2}{2\rho} + \frac{\rho}{2}$$ (21.12)

with the Ivantsov relation

$$\Delta = \sqrt{\pi P} e^P \mathrm{erfc}(\sqrt{P})$$ (21.13)

with the error function defined in (19.11) [83]. (See Exercise). For a small supercooling Δ, Peclet number P is small and the Ivantsov relation is approximated as

$$\Delta \approx \begin{cases} P(-\ln P - 0.5772 \cdots) & \text{for 3 dimensions} \\ \sqrt{\pi P} & \text{for 2 dimensions,} \end{cases}$$ (21.14)

and for high supercooling as $\Delta \to 1$, the Peclet number P is large and the Ivantsov relation is approximated as

$$\Delta \approx \begin{cases} 1 - 1/P & \text{for 3 dimensions} \\ 1 - 1/2P & \text{for 2 dimensions,} \end{cases}$$ (21.15)

Horvay and Cahn [83] extended the Ivantsov solution to the elliptic paraboloid:

$$z - Vt = \frac{1}{2R}(x^2 + a^2 y^2) + \frac{R}{2}$$ (21.16)

with the aspect ratio (ratio of x axis to y axis) a. The undercooling Δ and Peclet number $P = VR/2D$ is related as

$$\Delta = \frac{1}{a} P e^P \int_P^\infty \frac{e^{-w} dw}{\sqrt{w[w + (a^{-2} - 1)P]}} = \int_0^\infty \frac{Pe^{-w}}{\sqrt{(P + w)(P + a^2 w)}} dw.$$ (21.17)

For $a = 1$ the shape becomes the three-dimensional paraboloid of revolution and the Ivantsov relation (21.10) is reproduced. For $a = 0$ the system reduces to the two-dimensional problem, and Eq.(21.17) reduces to Eq.(21.13).

The Ivantsov solution indicates that the steady growth is possible in the form of a parabolic dendritic crystal. But the Ivantsov relation (21.10) shows that the growth velocity V and the tip radius R are not determined uniquely for a given undercooling Δ. It only determines the product $VR \sim P$. On the contrary, experiments show that for a given Δ, V and R are determined uniquely. There is some factor missing in the analysis by Ivantsov.

Exercise: Find the Ivantsov relation (21.13) for a two-dimensional dendrite, when the surface tension is absent.

Answer: Take the 2D parabolic coordinate

$$\begin{aligned}\xi &= r - z', & x &= \sqrt{\xi\eta} \\ \eta &= r + z', & z' &= \tfrac{1}{2}(\eta - \xi)\end{aligned} \tag{21.18}$$

with $r = \sqrt{x^2 + z'^2} = (\xi + \eta)/2$, the diffusion equation in steady state is written as

$$\sqrt{\eta}\frac{\partial}{\partial\eta}\left(\sqrt{\eta}\frac{\partial u}{\partial\eta}\right) + \sqrt{\xi}\frac{\partial}{\partial\xi}\left(\sqrt{\xi}\frac{\partial u}{\partial\xi}\right) + \frac{1}{l_D}\left(\eta\frac{\partial u}{\partial\eta} - \xi\frac{\partial u}{\partial\xi}\right) = 0. \tag{21.19}$$

Boundary conditions at the surface $\eta = \eta_i$ are expressed as the local equilibrium condition

$$u(\eta_i) = \Delta, \tag{21.20}$$

and the continuity condition

$$\eta_i + \xi\frac{\partial\eta_i}{\partial\xi} + \frac{\eta_i + \xi}{2v}\frac{\partial\eta_i}{\partial t} = -l_D\left(\eta_i\frac{\partial u}{\partial\eta} - \xi\frac{\partial\eta_i}{\partial\xi}\frac{\partial u}{\partial\xi}\right). \tag{21.21}$$

By taking the parabolic shape as $\eta_i = \rho$, the diffusion field u depends only on η from the symmetry and the diffusion equation becomes simple:

$$\frac{\partial}{\partial\eta}\left(\sqrt{\eta}\frac{\partial u}{\partial\eta}\right) + \frac{1}{l_D}(\sqrt{\eta}\frac{\partial u}{\partial\eta}) = 0. \tag{21.22}$$

The solution which satisfies the far field condition $u(\eta \to \infty) = 0$ and the local equilibrium condition (21.20) at $\eta = \rho$ is obtained as

$$u(\eta) = \Delta\left(1 - \frac{\int_\rho^\eta \eta^{-1/2}e^{\eta/l_D}\,d\eta}{\int_\rho^\infty \eta^{-1/2}e^{\eta/l_D}\,d\eta}\right). \tag{21.23}$$

The continuity boundary condition, (21.21), is written as

$$1 = -l_D\frac{\partial u}{\partial\eta} = \Delta\frac{\rho^{-1/2}e^{-\rho/l_D}}{\int_\rho^\infty \eta^{-1/2}e^{-\eta/l_D}\,d\eta}. \tag{21.24}$$

By changing the integration variable to $x = \sqrt{\eta/l_D}$ and using the Peclet number $P = \rho/l_D$, Eq.(21.24) is transformed to the desired result:

$$\Delta = 2\sqrt{P}e^P\int_{\sqrt{P}}^\infty e^{-x^2}\,dx = \sqrt{\pi P}e^P\,\mathrm{erfc}(\sqrt{P}) \tag{21.25}$$

with the error function (19.11).

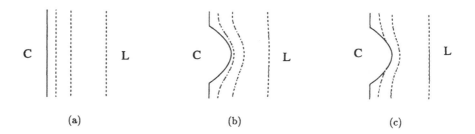

Figure 22.1: Isotherm distribution in the supercooled liquid around the interface (a) for a flat interface, (b) for a deformed interface without capillarity, (c) and for an interface with Gibbs-Thomson effect included.

22   Stability of a Flat Interface

In the section 19 we discussed the steady growth of a flat interface and found that for $\Delta < 1$, no steady state is possible. If the interface remains to be flat, the growth will slow down, and the growth velocity becomes zero even with a finite driving force. Another possible scenario of the time evolution of the flat interface is that it looses stability and deforms. We now study the latter possibility.

In order to discuss the extreme case of the diffusion limited growth, we assume hereafter an infinitely fast kinetics at the interface, $K \to \infty$, and the local equilibrium condition (18.11), or

$$u_i = \Delta - d\kappa. \tag{22.1}$$

First we give a qualitative explanation on the stability of the interface. If the flat interface moves steadily with a velocity V, the diffusion field u varies from Δ to 0 in a length scale of diffusion length $l_D = 2D/V$. The isotherm $u = const$ lies parallel to the interface $z' = z - Vt = 0$ as schematically shown in Fig.22.1a. When a part of the interface advances faster than the other by some fluctuation, what will happen to this pointed part? First we neglect the small effect of surface tension or capillarity. Then at the interface the diffusion field u_i takes a constant value Δ, irrespective of its deformation. Near the pointed part the isotherm $u_i = \Delta$ pushes forward other isotherms, and the isotherm density becomes high here, as shown in Fig.22.1b. The high density of isotherm means the steep slope of ∇u, and the heat flux $-D\nabla u$ increases and releases the latent heat quickly. The pointed part can grow faster than the undeformed part, and the deformation is enhanced: The flat interface becomes unstable. This instability is first studied analytically by Mullins and Sekerka, and is called Mullins-Sekerka instability [141, 122].

So far we considered the interface instability when the crystal is growing in the supercooled melt and the heat is transported through the liquid. If the crystal is supercooled and the melt is hot, what will happen for the interface deformation?

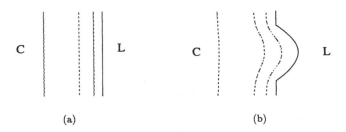

(a)   (b)

Figure 22.2: Isotherm distribution when the crystal is supercooled. (a) A flat interface, and (b) a deformed interface without capillarity.

Since the crystal is supercooled, the latent heat is released through the crystal. For a flat interface isotherms in the crystal are running parallel to the interface, as shown in Fig.22.2a. If a part of the crystal grows fast at some point by fluctuation, the separation between isotherms near the pointed part increases and the isotherm density decreases (Fig.22.2b). The slope of the isotherm decreases, and the latent heat is less effectively transported. Thus the growth rate there decreases, and the interface recovers the flat profile. In this case the interface is stable against fluctuation. There is an asymmetry and the instability occurs when the interface is propagating in the region where the transport is taking place.

We now consider the effect of surface tension on the stability of a flat interface. The surface tension lowers the degree of supercooling at the pointed part with positive curvature ($\kappa > 0$) as (22.1) and shown in Fig.22.1c. Since the driving force at the pointed part decreases, the growth rate there decreases. Thus the surface tension acts as a stabilizing factor for the flat interface. The total stability is determined by the competition between the diffusional destabilization and the energetic stabilization.

We now describe the stability in the linear analysis. The interface is assumed to be deformed sinusoidally with a wavelength $\lambda = 2\pi/q$ as

$$\zeta(x, y; t) = Vt + a_q \exp(\omega_q t) \cos(qx). \qquad (22.2)$$

The amplification rate ω_q determines the stability of the flat interface: When ω_q is negative, the interface is stable, whereas when ω_q is positive, it is unstable. Since the interface deformation influences the diffusion field u, it also has an additional variation to that of the flat interface in Eq.(19.4) as

$$u(x, y, z; t) = A \exp\left[-\frac{2(z - Vt)}{l_{\mathrm{D}}}\right] + \delta u_q \exp(\omega_q t) \cos(qx) \exp\left[-\Lambda_q(z - Vt)\right]. \qquad (22.3)$$

Here Λ_q should be positive, since the far field $u(z \to \infty)$ will not be affected by the interface deformation and the deviation should decay in z direction. By inserting

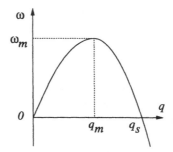

Figure 22.3: Dispersion relation of a sinusoidal deformation of a flat interface.

(22.3) in the diffusion equation (18.6) and by considering the terms containing $\cos qx$, one gets the relation

$$\frac{\omega_q}{D} + \frac{2\Lambda_q}{l_D} = -q^2 + \Lambda_q^2. \tag{22.4}$$

From the local equilibrium condition (22.1) with the curvature $\kappa \approx -\partial^2 z/\partial x^2$, A should be equal to Δ and the deformation amplitude δu_q and a_q are related as

$$\delta u_q = \left(\frac{2}{l_D}\Delta - d_0 q^2\right) a_q. \tag{22.5}$$

In the linear analysis the anisotropy in surface energy γ gives no contribution and d_0 denotes the isotropic part of the capillary length defined in (18.10). From the energy conservation or continuity equation (18.7), one gets the relation that $\Delta = 1$ and

$$\frac{\omega_q}{D} = -\left(\frac{2}{l_D}\right)^2 + \left(\frac{2}{l_D} - d_0 q^2\right)\Lambda_q. \tag{22.6}$$

From Eq.(22.4) and (22.6) the damping in z direction is determined as

$$\Lambda_q = \frac{2}{l_D} - \frac{1}{2}d_0 q^2 + |q|\sqrt{1 - \frac{2d_0}{l_D} + \frac{d_0^2 q^2}{4}}, \tag{22.7}$$

and the amplification rate as

$$\frac{\omega_q}{D} = \left(\frac{2}{l_D} - d_0 q^2\right)|q|\sqrt{1 - \frac{2d_0}{l_D} + \frac{d_0^2 q^2}{4} - \frac{3d_0 q^2}{l_D} + \frac{d_0^2 q^4}{2}} \tag{22.8}$$

which is depicted in Fig.22.3. For slow growth, the diffusion length $l_D = 2D/V$ is large; $l_D/d_0 \gg 1$. For a deformation with a wavelength less than l_D or $ql_D \gg 1$, the dispersion relation (22.8) is approximated as

$$\frac{\omega_q}{D} \approx (\frac{2}{l_D} - d_0 q^2)|q|. \tag{22.9}$$

For wavenumbers q larger than the stability value

$$q_s = \sqrt{\frac{2}{d_0 l_D}}, \tag{22.10}$$

the deformation damps down by the negative amplification rate ω_q. For the small wavenumbers q, however, the capillarity is not strong enough to stabilize the flat interface, and ω_q is positive. In terms of the wavelength, the mode with a modulation wavelength longer than the stability length

$$\lambda_s \equiv \frac{2\pi}{q_s} = 2\pi \sqrt{\frac{d_0 l_D}{2}} \tag{22.11}$$

is amplified. The most unstable mode has the wavenumber

$$q_m = q_s/\sqrt{3} \tag{22.12}$$

with the amplification rate

$$\omega_m = \frac{2}{3} V q_m \tag{22.13}$$

as is shown in Fig.22.3. So far we considered the one-sided model. If the diffusion constant in the crystal is the same with that in the liquid phase, namely in the symmetric model, the stability length is given by $\lambda_s = 2\pi\sqrt{d_0 l_D}$.

Exercise: Crystal in a spherical shape is growing in a static diffusion field $\nabla^2 u = 0$ with the local equilibrium boundary condition (18.11). Discuss its stability against the deformation as

$$r(\theta, \phi) = R(t) + a_{lm}(t) Y_{lm}(\theta, \phi) \tag{22.14}$$

with Y_{lm} a spherical harmonics in spherical coordinates (r, θ, ϕ) [140]. The Laplacian in spherical coordinate is expressed as

$$
\begin{aligned}
\nabla^2 &= \frac{1}{r^2}\frac{\partial}{\partial r}\left(r^2\frac{\partial}{\partial r}\right) + \frac{1}{r^2\sin\theta}\frac{\partial}{\partial\theta}\left(\sin\theta\frac{\partial}{\partial\theta}\right) + \frac{1}{r^2\sin\theta}\frac{\partial^2}{\partial\phi^2} \\
&= \frac{1}{r^2}\frac{\partial}{\partial r}\left(r^2\frac{\partial}{\partial r}\right) + \frac{1}{r^2}\hat{\Lambda}(\theta, \phi),
\end{aligned}
\tag{22.15}
$$

and spherical harmonics Y_{lm} are eigenfunctions of the angular part $\hat{\Lambda}$ with an eigenvalue $-l(l+1)$ as

$$\hat{\Lambda}Y_{lm}(\theta, \phi) = -l(l+1)Y_{lm}. \tag{22.16}$$

Also in the linear approximation the curvature $\kappa(\theta, \phi)$ can be represented as

$$\kappa(\theta, \phi) = \frac{2}{R} - \frac{1}{R^2}(l+2)(l-1)a_{lm}Y_{lm}(\theta, \phi). \tag{22.17}$$

Answer: The diffusion field is modified from Eq.(20.2) as

$$u(r,\theta,\phi) = \frac{A}{r} + \delta u_{lm}(r,t)Y_{lm}(\theta,\phi). \tag{22.18}$$

Since the field $u(r,\theta,\phi)$ should satisfy the diffusion equation in the static approximation $\nabla^2 u = 0$ or

$$\frac{\partial}{\partial r}\left(r^2\frac{\partial \delta u_{lm}}{\partial r}\right) - l(l+1)\delta u_{lm} = 0, \tag{22.19}$$

the deformation amplitude which decays far from the crystal is solved as

$$\delta u_{lm} = \frac{B}{r^{l+1}}. \tag{22.20}$$

The local equilibrium boundary condition (18.11) at the interface (22.14) is explicitly written as

$$u_i = A\left(\frac{1}{R} - \frac{a_{lm}}{R^2}Y_{lm}\right) + \frac{B}{R^{l+2}}Y_{lm} = \Delta - d\left[\frac{2}{R} - \frac{1}{R^2}(l+2)(l-1)a_{lm}Y_{lm}\right] \tag{22.21}$$

in the linear approximation, and two parameters A and B are determined as

$$A = R\left(\Delta - \frac{2d}{R}\right) \tag{22.22}$$

$$B = R^{l-1}\left[A - (l+2)(l-1)d\right]a_{lm} = R^l\left[\Delta - \frac{l(l+1)d}{R}\right]a_{lm}. \tag{22.23}$$

The continuity equation(18.7) is expressed as

$$V_n = \dot{R} + \dot{a}_{lm}Y_{lm}. \tag{22.24}$$

From the zeroth order, one gets the growth velocity

$$\dot{R} = D\frac{1}{R}\left(\Delta - \frac{2d}{R}\right) = \frac{D\Delta}{R}\left(1 - \frac{\rho_c}{R}\right). \tag{22.25}$$

From the linear order in a_{lm}, the modulation velocity is obtained as

$$\dot{a}_{lm} = \frac{D}{R^2}(l-1)\left[\Delta - \frac{d}{R}(l^2 + 3l + 4)\right]a_{lm} \equiv \omega_{lm}a_{lm}. \tag{22.26}$$

For small radius R when $\omega_{lm} < 0$, the deformation amplitude a_{lm} decays and the spherical shape is stable. For large radius when $\omega_{lm} > 0$, the deformation grows and the spherical shape becomes unstable. The critical radius for the l-th mode is given from the relation $\omega_{lm} = 0$ as

$$R_c(l) = \frac{d}{\Delta}(l^2 + 3l + 4) = \rho_c\frac{l^2 + 3l + 4}{2}. \tag{22.27}$$

Since the $l = 1$ model is marginal as $\omega_{1m} = 0$, the $l = 2$ mode is destabilized first at the radius R seven times of the critical nucleation radius $\rho_c = 2d/\Delta$. Thus a small sphere loses stability in the shape, and starts to deform.

Figure 23.1: Fractal dendrite of Au deposited on Ru [86].

23   Fractal Dendrite

If there is no surface tension ($d_0 = 0$), the dispersion relation (22.8) reduces to

$$\frac{\omega_q}{D} = \frac{2}{l_{\rm D}}|q|, \tag{23.1}$$

which is positive for all deformation wavenumbers q. The interface is always unstable, and it is most unstable for the largest $|q|$ or the finest structure of deformation. From Eq.(22.27) the spherical crystal is also unstable for any size if there is no capillarity effect. Therefore, one expects that the very fine and irregular structure will be realized for the crystal growing in the diffusion field without the surface tension. This kind of structure is actually observed in the vapor deposition on a very cold substrate, as shown in Fig.1.3a [41] and Fig.23.1 [86]. The adsorbed atom can diffuse randomly on the substrate, and its life time is very long on a cold substrate. In the meanwhile, the adatom collide with another adatom, coagulates and stop moving. Since the substrate is so cold that the once coagulated atoms never dissociate again to search for energetically more favorable position, the surface energy cannot play its role to stabilize the interface morphology. The irregular structure so grown is self-similar and called fractal, and is intensively studied recently [189, 10]. There are many other examples of fractal objects as aggregates grown by electrochemical deposition [130, 21] and bacteria colonies [132, 22]. We summarize the fractal theory relevant to crystal growth, and discuss fractal-to-compact crossover for the aggregate growing in a diffusion field of a finite density [191, 185].

23.1   Diffusion-limited Aggregation (DLA)

The irregular aggregate is first studied theoretically in the computer [197]. An immobile crystal particle is placed at some point as a seed, and a randomly diffusing gas particle is released far from the seed. When the gas particle comes into contact with the seed particle, it freezes and is incorporated into the aggregate. Then, the

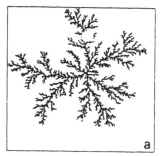 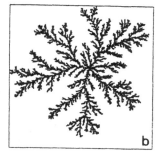 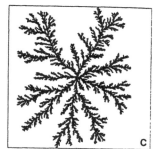

Figure 23.2: Diffusion-limited aggregation (DLA) grown in computer [150]. There are (a) 6×10^4, (b) 6×10^5 and (c) 6×10^6 particles in each cluster.

next gas particle is released far from the aggregate, and performs the random walk. After it makes contact with the aggregate, it freezes, and a new particle is released again. By iterating this procedure, the aggregate grows by incorporating atoms one by one. The grown aggregate is called the diffusion-limited aggregate (DLA), and its structure is shown in Fig.23.2. It is irregular, ramified and very open. The part looks similar to the whole by enlargement, and there is no characteristic length in the structure except the lower and upper cutoff lengths, the atomic and the system sizes. This self-similar object is called fractal. If there are $N(r)$ atoms in a region with a radius r, the number of atoms in the radius br is $N(br) = b^{D_f} N(r)$. By taking $b = r^{-1}$, then

$$N(r) = r^{D_f} N(1) \tag{23.2}$$

and D_f is called the fractal dimension. If the object fills the d dimensional space homogeneously with a finite density, $D_f = d$ and it is called compact. DLA is, on the contrary, fractal with $D_f = 1.71$ in $d = 2$ dimensions, and with $D_f = 2.49$ in $d = 3$ dimensions [134]. Since there are $N \sim r^{D_f}$ number of particles in a radius r, the aggregate density is

$$n(r) \sim \frac{N}{r^d} \sim r^{D_f - d}, \tag{23.3}$$

which becomes zero asymptotically for $r \to \infty$. It means that the aggregate has many open spaces. Zero asymptotic density is quite imaginable because the aggregate grows from a diffusion field of zero density: There is only a single gas particle during the whole growth simulation. Also a particle can walk forever until it comes to contact to the aggregate, and it means that the aggregate grows very slowly, $V \to 0$. The diffusion equation thus reduces to the Laplace equation $\nabla^2 u = 0$. The diffusion length $l_D = 2D/V$ is infinity, and the surrounding field as well as the aggregate have no characteristic length in the DLA growth.

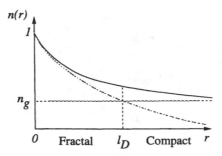

Figure 23.3: Schematic behavior of an aggregation density $n(r)$ growing in a diffusion field of a finite gas density n_g. The diffusion length l_{D} is the characteristic length scale of the crossover between the fractal and compact structure.

23.2   Fractal-to-Compact Crossover of an Aggregate in a Finite Gas Density

In a real experiment an aggregate grows from a gas with a finite density n_g, whose time evolution follows the diffusion equation. The growth rate of the aggregate V should no vanish, and the relation between the velocity V and the gas density n_{g} is of interest [185].

To realize the steady growth easily, we consider a unidirectional growth of an aggregate from a linear seed, similar to the "fractal forest" shown in Fig.1.3b. In front of the crystal, there are gas particles with an average density n_g and they are making random walks on a square lattice to the nearest neighbor site. When a gas particle comes into contact to a seed or to an aggregate grown from it, it freezes and becomes a member of the aggregate. The time scale is so chosen that in a unit time every gas particle makes one diffusional jump on average. By taking the lattice parameter as a unit of length, the diffusion constant D is 4. The aggregate front $h(t)$ is defined as the height of newly crystallized particle, and the velocity V is obtained from the time variation of $h(t)$. Since the gas particle is incorporated in the aggregate instantaneously and irreversibly, the density in front of the aggregate vanishes. It relaxes back to the given density n_g within the distance of diffusion length $l_{\mathrm{D}} = 2D/V$. Within this length, the gas density is low and the situation looks similar to the DLA growth: A gas particle sticks to the aggregate randomly and never melt back. Therefore, the grown aggregate has a fractal structure. But, since the aggregate grows from a gas with a finite density n_g, the material conservation does not allow the aggregate density to diminish as $n(r) \sim r^{D_{\mathrm{f}}-d}$ like the fractal object in the previous subsection does. For large distance $r \to \infty$, the aggregation density should saturate to n_g, as shown in Fig.23.3. A crossover of the aggregation density from the power law decay (23.3) to saturation takes place within a characteristic length l_{D}, and the

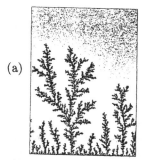 (a)   (b)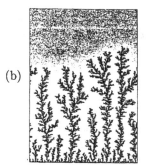

Figure 23.4: Irreversible and unidirectional solidification of aggregate from a gas with finite density: (a) $n_g = 0.08$, and (b) $n_g = 0.1$ [185].

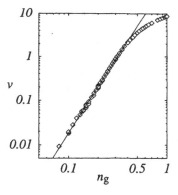

Figure 23.5: Velocity of the aggregate front v versus the gas density n_g, showing the power law dependence $v \sim n_g^{\nu}$ with $\nu = 1/(d - D_f)$ in $d = 2$ dimensions with $D_f = 1.71$ [185].

relation

$$l_D^{-(d-D_f)} \sim n_g \tag{23.4}$$

is expected. There should be a crossover from the fractal to compact structure at the length scale of order l_D. This gives the relation between the aggregate growth velocity V and the gas density n_g as

$$V \sim l_D^{-1} \sim n_g^{1/(d-D_f)}. \tag{23.5}$$

In two dimensional system ($d = 2$), Monte Carlo simulation shows that the aggregate consists of many branches of irregular dendrites, and the separation between branches decreases as the gas density n_g increases, as shown in Fig.23.4. This is compatible with the decrease of the characteristic length as n_g increases. The growth velocity of the aggregate V is found to increase in powers of the gas density n_g with the exponent compatible with the fractal dimension $D_f = 1.71$ in two dimensions, as shown in Fig.23.5. The scaling (23.5) is recently confirmed in the deposition experiment of silver metal leaves from $AgNO_3$ solution [139].

24   Capillary Effect and Regular Dendrite

Without the surface tension, the crystal profile is shown to take a random and irregular form. In reality, most metals and some plastic crystals growing in a free and open space takes a regular dendritic form with a stable tip oriented in a special crystallographic direction, as shown in Fig.24.1 [84]. The tip of the dendrite is parabolic and the tip radius R and the growth velocity V satisfies the Ivantsov relation (21.10) with the undercooling Δ (Fig.24.2) [84]. In experiments the undercooling Δ determines R and V uniquely, contrary to the Ivantsov solution where an infinite degeneracy is expected. Capillarity may select a unique operating point among multitude of possibilities. It introduces a new characteristic length, the capillary length $d = \tilde{\gamma} C_p T_M / L^2$ of (18.10). It is intrinsic, since it is determined only by the material parameters and every other length as ρ or l_D is expected to be scaled with d.

Since the main interest lies in the interface structure, it seems advantageous to integrate out the diffusion field and to derive the interface dynamics. Many local models for the evolution of the interface profile were proposed, and gave contributions to understand the problem of the velocity selection [40, 17, 18, 100, 101]. But the true interface dynamics is nonlocal even in quasi-stationary approximation, as explained in detail in Appendix A24.1. We summarize the results of recent studies on the nonlocal model.

24.1   Velocity Selection by Microscopic Solvability

Due to mathematical simplicity, the analysis is mainly done on two dimensional dendrite [124, 105, 33, 156]. The capillary length is defined in terms of the two-dimensional surface stiffness $\tilde{\beta}$. The surface tension β is assumed to have four-fold rotational symmetry reflecting the crystalline order of, such as, succinonitrile.

$$\beta(\theta) = \beta_0(1 + \tilde{\epsilon}\cos 4\theta). \tag{24.1}$$

For positive $\tilde{\epsilon}$, $\theta = 0$, $\pm\pi/2$, and π corresponds to the maximum of the surface tension, and to the corners in equilibrium shape (Fig.6.5). The surface stiffness is expressed as

$$\tilde{\beta} = \beta + \frac{\partial^2 \beta}{\partial \theta^2} = \beta_0(1 - 15\tilde{\epsilon}\cos 4\theta), \tag{24.2}$$

and the capillary length (18.10) is then written as

$$d = d_0(1 - \epsilon\cos 4\theta) \tag{24.3}$$

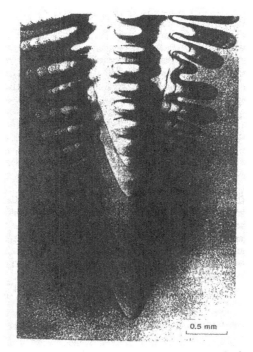

Figure 24.1: Dendrite of succinonitrile [84].

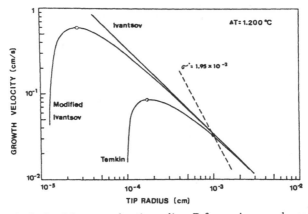

Figure 24.2: Velocity V versus the tip radius R for a given undercooling Δ [84].

with d_0 being the magnitude of the capillarity length and $\epsilon = 15\tilde{\epsilon}$ the strength of the anisotropy. Orientations $\theta = 0, \pm \pi/2, \pi$ correspond to the minimum of the stiffness and the capillary length.

By including the capillarity, the stability length appears, for example, $\lambda_s = 2\pi\sqrt{d_0 l_D/2}$ in the one-sided model. When the tip radius ρ is smaller than λ_s, the tip is unable to grow further due to the capillary effect, and the interface turns to be flat. When ρ is larger than λ_s, the tip is unstable to the deformation due to the Mullins-Sekerka instability. Therefore, the dendrite tip radius is expected to be of order λ_s. Then, one introduce a dimensionless parameter

$$\sigma \equiv \frac{d_0 l_D}{\rho^2} = \frac{d_0}{\rho P} = \frac{2Dd_0}{v\rho^2} = \frac{d_0 v}{2DP^2}, \qquad (24.4)$$

which is called the stability parameter. Since the capillary length d_0 is small, this σ is small, and one may try the expansion of the profile in powers of σ. However, as the capillary effect is contained in the Gibbs-Thomson condition, it is coupled to the curvature κ, the highest derivative of the interface profile as $\sigma\kappa \sim -\sigma\partial^2 z/\partial x^2$. In this case, normal perturbation fails and one has to use a singular perturbation [16]. There are many theoretical works on the dendrite theory which are summarized in many reviews [124, 105, 33, 156].

In solving the nonlocal interface equation, boundary conditions has to be considered appropriately. Far down the dendrite tail, the curvature is so small and the profile should approach to the Ivantsov parabola. By integrating the interface profile from the tail to the tip, one gets in general a finite slope Θ at the tip and the solution cannot be extended symmetrically to the other side of the tail. Slope at the tip can vanish only for special choice of the growth velocity v or the stability parameter σ. This is called the solvability condition [124, 105, 33, 156], and is sketched in a little more detail in Appendix A24.2. Here I summarize the main conclusions.

1. Without surface anisotropy $\epsilon = 0$, there is no steady state for the dendritic growth.

2. With an anisotropy $\epsilon \neq 0$, the symmetric dendrite grows steadily with its tip oriented in the direction of the stiffness minimum $\theta = 0$.

3. The stability parameter σ is found to depend only on an anisotropy parameter ϵ, but is independent of the undercooling Δ. For small ϵ, σ is approximately proportional to $\epsilon^{7/4}$ and thus $v\rho^2$ remains constant for various undercooling Δ:

$$v\rho^2 = 2Dd_0/\sigma(\epsilon) \propto \epsilon^{-7/4}. \qquad (24.5)$$

The value of $\sigma(\epsilon)$ for the one-sided model is shown to be twice as large as that of the symmetric model [138]. By combining with the Ivantsov relation, $v\rho = 2DP(\Delta)$, the growth rate is obtained as $v \approx P^2$. For $d = 2$ it reads as $v \approx \Delta^4$ at small undercooling where the approximation (21.14) holds.

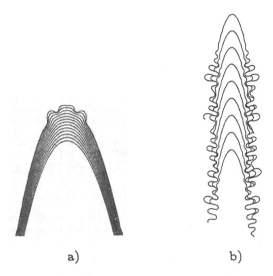

Figure 24.3: Simulation of the dynamical evolution of the dendritic crystal at (a)
$\varepsilon = 0$ (isotropic surface energy), and $\varepsilon = 0.10$. The isotropic surface energy (a) shows
the tip splitting, and the anisotropic one grows stably in (b) [165].

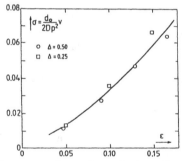

Figure 24.4: The stability parameter σ dependent on the anisotropy parameter ε at
two different undercoolings $\Delta = 0.25$ and $\Delta = 0.50$ [165].

This result is confirmed by the numerical simulation by solving the diffusion equation in a quasi-stationary approximation by the boundary element method and integrating the shape evolution [165], as explained in Appendix A24.3: For $\epsilon = 0$, the
dendrite tip splits as it develops in time (Fig.24.3a), and for $\epsilon > 0$ the tip stable dendrite grows as shown in Fig.24.3b. The parameter $\sigma = d_0 v / 2DP^2$ is found to depend
on anisotropy ϵ but is independent of the supercooling Δ, as shown in Fig.24.4.

In the three-dimensional experiment of succinonitrile dendrite growing in its melt

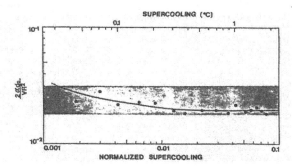

Figure 24.5: The stability parameter σ at various undercooling Δ's for succinonitrile [84].

(Fig.24.1) [84], the relation $VR^2 = const$ is found to be satisfied as shown in Fig.24.2 and Fig.24.5. The strong anisotropy dependence of VR^2 is, however, not observed experimentally. There is some argument that anisotropy is not necessary in the mode selection [71], but at least it seems plausible that the dendrite grows in the direction of the stiffness minimum, since the minimum stiffness means less effective in suppressing deformation and the deformation grows fastest in that direction. If there is no preference in orientation, the growing tip is very susceptible to the orientation fluctuation, and may split or form irregular dendrite.

The chaotic behavior of the isotropic interface is already found in the Kuramoto-Sivashinsky equation discussed in section 17. There is another example of tip instability in an isotropic system, a viscous finger problem [162, 152]: A viscous fluid is confined in a narrow gap between two glass plates of a Hele-Shaw cell [75]. When the fluid with low viscosity (like air) is pushed into the fluid with high viscosity (like water), the meniscus between two fluids is unstable and forms finger-like pattern, known as the viscous finger. The problem is formulated similar to the dendritic growth. One essential difference between the two problems is that the surface tension is isotropic in the viscous finger problem, whereas it is anisotropic in the crystal growth. The pattern realized in the radial Hele-Shaw cell shows a branched ramified structure as shown in Fig.24.6 [152]. By engraving a line groove in some portion on one glass plate of a Hele-Shaw cell, the tip stability is realized as shown in Fig.24.7a [131]. The fluid dendrite thus obtained satisfies the relation $v\rho^2 = constant$ as shown in Fig.24.7b, until the tip radius ρ becomes too small. This experiment clearly shows the importance of the anisotropy in stabilizing the dendrite tip [19].

So far, theories and simulations dealt with two-dimensional(2D) dendrite growth. Recently, solvability condition is successfully applied to the three-dimensional (3D) dendrite [14, 36]. The surface free energy $\gamma(\theta, \phi)$ now has an orientation dependence on two Euler angles, θ and ϕ of the normal vector to the surface. In the theory, a cubic anisotropy

$$\gamma(\theta, \phi) = 1 + \tilde{\epsilon}(4\cos^4\theta + 3\sin^4\theta + \sin^4\theta \cos 4\phi) \qquad (24.6)$$

Figure 24.6: Viscous finger pattern in a radial Hele-Shaw cell [152].

(a)

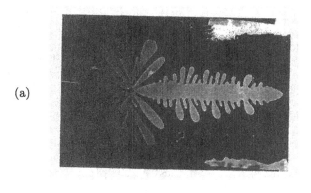

(b)

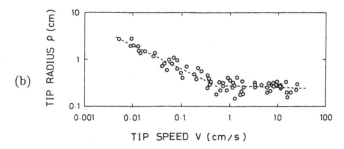

Figure 24.7: (a) Viscous finger with a groove line cut on one glass plate. Along the groove grows a tip-stable parabolic dendrite, whereas other tips are unstable to splitting. (b) Log-log plot of the tip radius of the fluid dendrite versus tip speed [131].

is assumed. The selection problem is solved in two stages: Near the dendrite tip, the shape is perturbed as

$$z = -\frac{r^2}{2R} + A_4 r^4 \cos 4\phi + \cdots \tag{24.7}$$

with $r^2 = x^2 + y^2$. The growth rate V and the tip radius R is found to satisfy the same relations as in the 2D case: the Ivantsov relation

$$VR = 2DP(\Delta) \sim -\frac{\Delta}{\ln \Delta}, \tag{24.8}$$

for small supercooling Δ, and the solvability scaling relation

$$VR^2 = \frac{2Dd_0}{\sigma(\epsilon)} \sim \epsilon^{-7/4}. \tag{24.9}$$

Therefore, the anisotropy dependence of the velocity V and the tip radius R is expected to be the same with the 2D case [14].

The analysis near the tip cannot be extended to the dendrite tail straightforwardly, since the correction $A_4 r^4 \cos 4\phi$ grows faster than the unperturbed Ivantsov paraboloid $-r^2/2R$ [36]. But the correction form indicates that four fins extend in the directions, $\phi = 0$, $\pi/2$, π, $3\pi/2$. When one looks far down the dendrite, the crystal interface is almost parallel to z axis, and the diffusion field hardly varies in z direction; $\partial^2 u/\partial z^2 \approx 0$. Then the evolution of the dendrite fin in the tail region is essentially controlled by the two dimensional diffusion [36]. The four fins that have started to grow near the tip develop into two-dimensional parabolic dendrites in xy cross section. For example, down at a height $z(< 0)$ from the tip, a dendrite growing in x direction takes a form

$$x = |z|\frac{v}{V} - \frac{y^2}{2\rho}, \tag{24.10}$$

where ρ and v are the selected values of radius and velocity of the 2D dendrite. Here the time t is replaced by the height $|z|$ divided by the 3D growth velocity V, since the profile at a height z develops to that at a height $z - Vdt$ after a time dt in the steady state

Between the two extreme asymptotics, Eqs. (24.7) and (24.10), an intermediate asymptotics is found [36] until the arm length exceeds the 2D diffusion length. In this intermediate region four fins affect their growth mutually. At a height $|z|$ from the tip, the total cross-sectional area of fins should be that of the Ivantsov paraboloid for the steady growth. If the fin protrudes a distance x with a width y, then the area is about $xy \sim |z|$. Even though x and y depends on the time $|z|$, the dimensionless parameter of the system

$$\sigma_2 = \frac{v\rho^2}{2Dd_0} \sim \frac{dx}{d|z|} \Big/ \left(\frac{d^2x}{dy^2}\right)^2 \tag{24.11}$$

may remain constant [5]. By assuming the profile $x \sim |z|^\alpha$ and $y \sim |z|^{1-\alpha}$, σ_2 is calculated as $\sigma_2 \sim |z|^{3-5\alpha}$. If σ_2 remains constant, then $\alpha = 3/5$, and the arm length

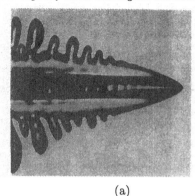
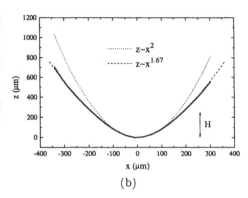

(a)   (b)

Figure 24.8: (a) Xenon dendrite and (b) tip contour compared with a power law fit to parabola and $z \sim x^{1.67}$ [26].

x grows as $x \sim |z|^{3/5}$ and the arm width y increases as $y \sim |z|^{2/5}$. This behavior is originally found in the two-dimensional Hele-Shaw flow problem numerically and analytically [5], and later applied to the dendritic growth [36]. The profile of the fin in xz cross-section is written as $z \sim x^{5/3}$ in this intermediate region. The profile is recently observed in the experiment of Xenon dendrite as shown in Fig.24.8 [26]. When the arm length exceeds the 2D diffusion length, the final aymptotics sets in and the arm length grows with a constant velocity. For the details, see the references [14, 36].

24.2 Tip Stability and Sidebranches

With an anisotropy ϵ in the surface tension, the parabolic crystal with a tip radius R is found to grow steadily with a velocity V in the supercooled melt with the supercooling Δ. V and R are uniquely determined from both the Ivantsov relation $VR = 2DP(\Delta)$ and the solvability condition $VR^2 = 2Dd_0\sigma^{-1}(\epsilon)$. However, the dendrite is growing in the environment with fluctuations, for instance, created by thermal noise or by hydrodynamic convections. Is the tip stable against these fluctuations? Another problem is the origin of the sidebranches. According to the solvability theory, the tip of the selected dendrite is almost parabolic without sidebranches. How, then, are the sidebranches created in experiments?

If the crystallizing front is flat, Mullis-Sekerka instability takes place and the front is unstable. It is most unstable against the fluctuation with the wavenumber $q_m = \sqrt{2/(3d_0l_D)}$ with the exponential amplification of the amplitude with the rate $\omega_m = 2vq_m/3$, as explained in the section 22. But actually the dendrite tip is curved parabolic as

$$z = \zeta(x) = vt - \frac{x^2}{2\rho} \tag{24.12}$$

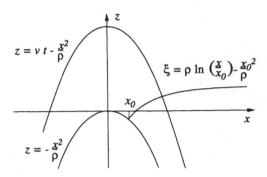

Figure 24.9: Trajectory $z = \xi(x)$ of the deformation on the interface, which is alway perpendicular to the interface profile, $z = \zeta(x)$.

in two dimensions. Since the interface deformation grows in the normal direction $\mathbf{n} = (n_x, n_z) \propto (-d\zeta/dx, 1)$ as shown in Fig.24.9, the node of deformation follows the trajectory $z = \xi(x)$

$$\frac{d\xi}{dx} = \frac{n_z}{n_x} = -\left(\frac{d\zeta}{dx}\right)^{-1} = \frac{\rho}{x} \tag{24.13}$$

or, by integration,

$$\xi(x) = \rho \ln x + C. \tag{24.14}$$

When the initial deformation at $t = 0$ is given at position $(x_0, -x_0^2/2\rho)$ on the interface, the integration constant C is determined as $C = -x_0^2/2\rho - \rho \ln x_0$. At time t, the deformation has been propagated at the site $(x, \zeta(x))$ determined by crossing point of $\zeta(x)$ and $\xi(x)$ as

$$\zeta(x) = vt - \frac{x^2}{2\rho} = \rho \ln \frac{x}{x_0} - \frac{x_0^2}{2\rho}. \tag{24.15}$$

Asymptotically for large t, x increases as $\sqrt{2\rho vt}$, but its height $z = \zeta(x)$ increases slowly as $\rho/2 \cdot \ln t$. The vertical separation between the dendrite tip at $(0, vt)$ and the deformed position at $\approx (\sqrt{2\rho vt}, \rho/2 \ln t)$ increases as vt. Though the deformation amplifies with the rate ω_m, it is convected down with a velocity v and the tip remains stable. This is called the convective stability of the dendrite tip. Down convected noise is amplified to form sidebranches. If this scenario is correct and sidebranches are originated from the random noise at the tip, sidebranches on the different side of the dendrite are expected to be uncorrelated. By measuring the correlation the noise origin of sidebranches is confirmed [54]. The external oscillatory flow imposes systematic fluctuation at the dendrite tip, and the synchronized formation of sidebranches is observed [28]. Of course, the tip stability depends on the amplitude of the initial noise. If the initial noise is too strong, the deformation is strongly amplified.

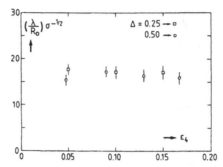

Figure 24.10: The sidebranch periodicity λ near the dendrite tip is proportional to the wavelength of the most unstable mode $\lambda_m \sim \rho\sqrt{\sigma}$, irrespective of the undercooling Δ and the anisotropy ε [165].

When the deformation amplitude becomes larger than the tip radius ρ before the deformation is convected down the length of order ρ, the dendrite tip splits [37]. In fact, the fractal structure is realized for NH_4Cl dendrite growing in a Hele-Shaw cell when a bottom glass is cut rough randomly [82].

We now study the sidebranch formation more quantitatively. The convected deformation amplifies and becomes a sidebranch with a periodicity λ about that of the most unstable mode,

$$\lambda_m = \frac{2\pi}{q_m} = 2\pi\sqrt{\frac{3}{2}}\sqrt{d_0 l_D} = 2\pi\sqrt{\frac{3}{2}}\rho\sqrt{\sigma}. \tag{24.16}$$

In the numerical simulation, the wavelength λ of the sidebranch is defined by dividing the growth rate v by the number of sidebranches produced per unit time. The ratio of the sidebranch periodicity λ to $\rho\sqrt{\sigma}$ was fount to be independent of the supercooling Δ and the anisotropy ϵ as shown in Fig.24.10 [165].

As the perturbation slides down along the side of the dendrite $z = -x^2/2\rho$, the normal velocity v_n decreases as $v_n = -v(\partial\zeta/\partial x)^{-1} \sim v\sqrt{\rho/|z|}$, as shown in Fig.21.1. Then the local diffusion length increases as $l_D(z) = 2D/v_n(z) \sim l_D\sqrt{|z|/\rho}$ as well as the most unstable wavelength

$$\lambda_m(z) \approx \lambda_m(0)\left(\frac{2|z|}{\rho}\right)^{1/4} \sim l_D(z)^{1/2}. \tag{24.17}$$

The periodicity is expected to increase as the sidebranching is convected down the shaft of the dendrite. Along with the variation of the periodicity $\lambda_m(z)$, the amplification rate $\omega_m(z) = 2v_n(z)q_m(z)/3$ of the most unstable mode varies while the

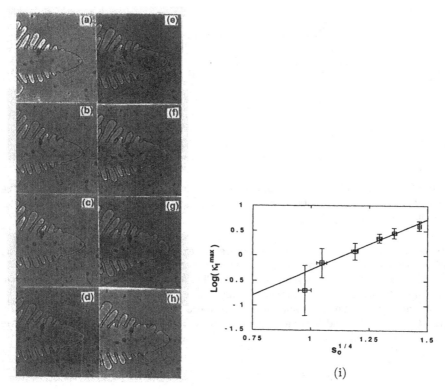

(i)

Figure 24.11: (a-h) The time sequence of the disturbance amplification near the tip of the dendrite to the sidebranch. (i) The maximum curvature κ_1 versus the arc length s [157]. κ_1 is proportional to the noise amplification.

perturbation is convected down.

$$\omega_m(z) = \frac{2v_n(z)q_m(z)}{3} \approx \frac{v}{\lambda_m(0)}\left(\frac{\rho}{2|z|}\right)^{3/4}. \tag{24.18}$$

Therefore, when the perturbation is convected down to a height $|z|$ during a time $t = |z|/v$, the deformation is totally amplified by a factor A;

$$A \sim \exp\left[\int_0^t \omega_m(z)dt\right] \sim \exp\left[\frac{\rho^{3/4}}{\lambda_m(0)}z^{1/4}\right] \sim \exp\left[\frac{(2/3)^{1/4}}{2\pi}\sigma^{-1/2}\left(\frac{|z|}{\rho}\right)^{1/4}\right]. \tag{24.19}$$

Precise expressions of A are obtained for the two-dimensional dendrites [155, 11], and for the three-dimensional axisymmetric dendrite [123]. If the initial perturbation is the thermal noise at the dendrite tip, it is found to be too small to explain the observed position of the **first** sidebranch. On applying a noise near the tip by a localized heat pulse, the noise amplification to the sidebranch (24.19) is studied experimentally [157]. The maximum of the curvature on the sidebranch is found to increase as $\kappa_1 \sim \exp(as^{1/4})$ during the noise is convected down the dendrite till the arc length s, as shown in Fig.24.11.

As explained in the previous subsection, there is the latest theory of the three-dimensional non-axisymmetric dendrite with four fins formed a little down the tip [36]. Since the profile of the fin $z = \zeta(x) \sim x^{5/3}$ has a slope as $\partial\zeta/\partial x \sim z^{2/5}$ which is milder than that of the parabolic dendrite $\partial\zeta_{IV}/\partial x \sim z^{1/2}$, the convective effect is small and the dendrite is more susceptible to the noise. At the height $|z|$, the most unstable mode has the short wavelength $\lambda_m(z) \sim z^{1/5}$, the large amplification rate $\omega_m(z) \sim z^{-3/5}$, and the large total amplification factor $A \sim \exp\left[C\sigma^{-1/2}z^{2/5}\right]$. The more precise formula is given by Brener and Temkin [38]. The amplification factor A increases steeply along the fin compared to the parabolic case, (24.19). The first side branch is in fact observed on the fin of the Xenon dendrite at the position expected from the thermal initial noise [26].

24.3 Dendrite in a Channel

For the case of a free dendrite, the anisotropy of the surface energy is shown to stabilize the dendrite tip. When the crystal grows in a channel with non permeable walls, interaction with the channel walls through the diffusion field allows stable stationary patterns even for a system without anisotropy [104]. The wall provides the effective anisotropy in the system and the steady growth of a symmetric finger-shaped crystal is possible if the dimensionless supercooling is large, $\Delta > \frac{1}{2}$ [31]. Due to the global conservation of the energy, the width w of a symmetric finger in a channel of width λ should be equal to $\Delta\lambda$; $w = \Delta\lambda$. If Δ is small, the symmetric finger is far from the wall and it cannot maintain the stable form. Therefore, Δ should be larger than 1/2. For a fixed supercooling Δ, the growth velocity v varies as a function of the channel width λ as shown in Fig.24.12. The result is obtained by

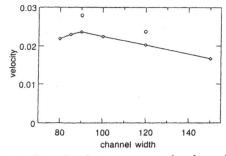

Figure 24.12: Growth velocity v versus the channel width λ [31, 35].

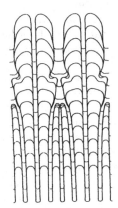

Figure 24.13: Asymmetric finger along the wall [35].

the extension of the numerical simulation described in Appendix A24.3. Velocity v has a maximum, and it decreases for a large channel width λ. For an infinitely wide system, as we discussed section 24.1, the isotropic dendrite cannot grow steadily with a finite velocity. When the width becomes too wide, the finger splits to have shorter period. Solvability theory by Brener, Geilikman and Temkin [31] found the maximum velocity $v \sim (D/d_0)(\Delta - \frac{1}{2})^{7/2}$ at the channel width $\lambda \sim d_0(\Delta - \frac{1}{2})^{-5/2}$. It is natural to assume that the state with a maximum velocity is selected and realized.

In the simulation, however, the asymmetric dendrite is observed which grows along the wall, as shown in Fig.24.13 [35]. In the simulation a mirror boundary condition is used: $u(x, y) = u(-x, y) = u(\lambda - x, y)$ for the diffusion field and $\zeta(x, y) = \zeta(-x, y) = \zeta(\lambda - x, y)$ for the interfacial profile with $0 < x < \lambda/2$. This asymmetric dendrite along the wall thus really means a double-finger structure with its mirror image at negative x. This structure is called "doublon" for short [87]. In a time-dependent simulation model [87], the doublon is also found in the system with periodic boundary conditions. If two fingers are growing side by side, one generally anticipate that the one which steps little ahead by fluctuation wins the competition against the other by gaining more diffusion field supply. In the present situation, when one wins, the periodicity λ increases and the growth rate v decreases according to Fig.24.12. The one ahead is caught up by the nearby finger. Thus the doublon can survive as a stable profile even at $\Delta < \frac{1}{2}$. Recent solvability analysis shows that the doublon grows with the velocity v proportional to $v \sim (D/d_0)\Delta^9$ without an anisotropy [15].

24.4 Morphology Diagram

From the findings in the previous subsections, an isotropic crystal (if it exists) can grow radially outward with a doublon structure in an open space. Since the separation

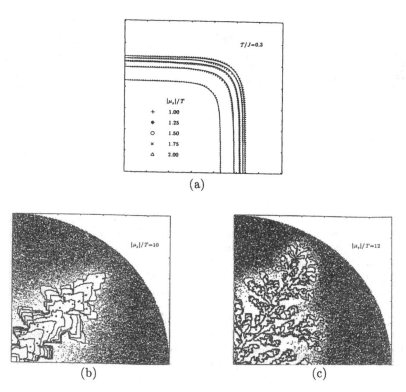

Figure 24.14: (a) Equilibrium shape of a crystal coexisting with lattice gas atoms in a closed system. Simulation results (symbols) for different $\Delta\mu$ are compared with the theoretical results drawn by curves. Time evolution of a crystal in an open system are depicted at different driving forces; (b) dendritic shape at $\Delta\mu/k_BT = 10$, and (c) irregular structure with tip splittings at $\Delta\mu/k_BT = 12$ [166].

between two double-fingers widens as they grow, the space should be filled by new fingers developed from sidebranches. The structure as a whole should have a spherical and convex envelope, and is called the compact seaweed (CS) structure [34] or dense branched morphology(DBM) [20]. With an anisotropy, dendrites grow in directions of stiffness minimum. When the dendrite tips are separated radially, the secondary arms from the sidebranches fill the space with a concave envelope. This is called the compact dendrite (CD) [34].

The morphological transition from CD to CS pattern is first obtained in the Monte Carlo simulation of a crystal growth from a lattice gas [166], as is shown in Fig.24.14. The growth is similar to the DLA growth from a finite density gas, but with an

Figure 24.15: Morphological transition from (a) dendrite, (b) intermediate, and (c) dense branched morphology (DBM) of a hexatic liquid crystal [151]. Undercoolings are (a) $\Delta = 0.24$, (b) 0.45 and (c) 0.63.

interfacial energy included. The interaction is the same with the Ising model (6.33) such that a broken bond from a crystal atom costs an energy J. When the gas atom diffuses and touches the crystal interface, it tries crystallization. If it crystallizes, there is a chemical potential gain $-\Delta\mu$ but has to pay an energy cost $(z/2 - n)J$ if the crystallized site has n nearest neighbor crystal atoms among z coordination number. These energy changes should be taken into account in the Boltzmann weight as is described in the subsection 9.1. One also has to consider the melting process from the crystal surface to satisfy the detailed balance.

In a closed system where the total number of gas and crystal atoms is fixed, an equilibrium shape is realized as is shown in Fig.24.14a. In an open system, the gas atom is fed from a particle reservoir with a fixed density at a distance far from the growing crystal. Then the crystal grows steadily. For small $\Delta\mu$ as $\Delta\mu/k_BT = 10$, the crystal grows in diagonal [11] direction in a regular dendritic form, as shown in Fig.24.14b. At a larger $\Delta\mu$ as $\Delta\mu/k_BT = 12$, the crystal grows in an irregular form with concave envelope, as shown in Fig.24.14c. Thus simulation clearly shows a crossover in the growth morphology. There are many similar simulation works on the morphological transitions [175, 176]. There is also an experimental observation of the morphological crossover in the growth of columnar hexagonal crystal, as shown in Fig.24.15 [151].

Analytical studies are also performed on the dynamical selection of morphology [34]. It is natural to assume a maximum velocity criterion that the pattern with maximum velocity is selected. With this hypothesis, Brener, Temkin and Müller-Krumbhaar have derived the morphological phase diagram in the phase space of the supercooling Δ and the anisotropy ϵ, as shown in Fig.24.16 [34]. Since they didn't considered doublons in an isotropic or weakly anisotropic region, CS structure can appear only for $\Delta > 1/2$. By considering the double-finger structure, the crossover between the CS and CD structure is expected to take place at $\Delta \sim \epsilon^{7/20}$.

At small undercooling Δ and small anisotropy ϵ, the noise is expected to be important and induce tip splitting. When the tip is destroyed, the structure becomes

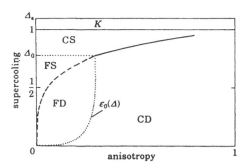

Figure 24.16: Morphological phase diagram in the phase space of the undercooling Δ and the anisotropy strength ε. CS: compact seaweed, equivalent to DBM, CD: compact dendrite, FS: fractal seaweed, and FD: fractal dendrite [34].

fractal for a short length scales, as is described in the subsection 23.2. These structures are called fractal seaweed (FS) and fractal dendrite (FD) in their morphology diagram, Fig.24.16. For the details, refer the original paper [34].

25 Unidirectional Solidification from the Solution

An example in which the growth pattern is controlled by the material diffusion is the alloy crystal growth from solution. Solution of a binary alloy is encapsulate in a thin Hele-Shaw cell, a thin rectangular parallelepiped cell (Fig.25.1). One end of the cell is kept hot and the other end is kept cold to impose a temperature gradient

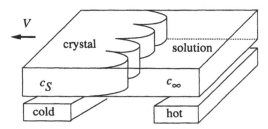

Figure 25.1: Unidirectional crystal growth from solution in a Hele-Shaw cell. Solution is placed in a narrow gap between two glass plates, heated at one end and cooled at the other end. The cell is pulled steadily in the cool region, facilitating the steady growth of the crystal.

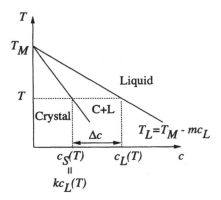

Figure 25.2: Equilibrium phase diagram of a binary alloy. The liquidus temperature T_L decreases from the melting temperature T_M of the pure A material by increasing the concentration c of B atoms.

over the sample. When the cell is pulled in the cold side, the solution is crystallized unidirectionally. The crystal-liquid interface lies normal to the pulling direction.

Since the liquid is hot and the crystal is cold, the latent heat is released in the cold crystal. In this case, the heat diffusion will not induce instability of the flat interface as explained in the section 22. But there is a slow material transport which controls the crystal growth. The solubilities in the liquid and crystal are different, as is shown in the phase diagram Fig.25.2. The component less soluble in the crystal is expelled in the liquid and should be transported far away down by the diffusion. The other component, more soluble in the crystal should be supplied far away from the liquid by the diffusion. This material diffusion causes the similar interface instability as the heat transport did in the melt growth, and leads pattern formation with periodic structure. To summarize, in the unidirectional growth heat conduction stabilizes the interface and the chemical diffusion causes the instability.

Since the heat conduction is faster than the material diffusion, and the heat conductivity in the crystal and liquid are almost the same, one often uses the approximation that the thermal gradient G_T is the same in both phases. When the sample is sandwiched in a thin cell, the heat conduction is mediated by the cell wall, and the thermal conductivities in both crystal and liquid phases are the same. Then the temperature distribution is simply described as

$$T(z) = T_0 + G_T z. \tag{25.1}$$

25.1 Fundamental Equations

We consider the phase diagram of AB alloy as shown in Fig.25.2. By including B

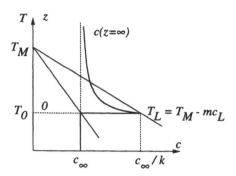

Figure 25.3: Concentration distribution drawn in a heavy line at the vertical axis representing the temperature T and the height z simultaneously from the relation $T = T_0 + G_T z$. Here $T_0 = T_M - mc_\infty/k$.

atoms with a concentration c, the melting temperature decreases from that of the pure A atoms, T_M. At a temperature T below T_M, the crystal phase is stable when the concentration c is less than $c_S(T)$, and the liquid phase is stable when c is more than $c_L(T)$. Between these two concentration, $c_S(T) < c < c_L(T)$, the crystal and liquid phases coexists. Phase separation between the crystal and the liquid phases takes place, and a crystal with a concentration $c_S(T)$ coexists with liquid with a concentration $c_L(T)$. The liquidus line is well approximated by a line

$$T_L = T_M - mc_L(T) \tag{25.2}$$

with a slope m. The partition (or segregation) coefficient k is defined as the concentration ratio between the coexisting two phases as

$$k = \frac{c_S(T)}{c_L(T)}, \tag{25.3}$$

and the miscibility gap is defined as

$$\Delta c = c_L(T) - c_S(T). \tag{25.4}$$

By pulling the two-dimensional cell with the solution of concentration c_∞ to the low temperature side with a constant velocity V, the steady crystallization takes place. From the global conservation of material, the concentration of the crystal alloy should also be c_∞, as denoted in the equilibrium phase diagram, Fig.25.3. First we consider the case when the surface is flat at $z = 0$. Since the material diffusion in the crystal is negligible, the concentration in the crystal should be $c_\infty = c_S^0$. Assuming that the local equilibrium is satisfied at the interface, the liquid concentration at the interface is $c_L^0 = c_\infty/k$ and an interface temperature is

$$T_0 = T_M - \frac{mc_\infty}{k}. \tag{25.5}$$

A concentration difference is created in the liquid at the interface c_L^0 and far from there c_∞. The concentration variation follows the diffusion equation

$$\frac{\partial c}{\partial t} = V\frac{\partial c}{\partial z} + D_c\nabla^2 c, \qquad (25.6)$$

with D_c being the chemical diffusion constant, and the first term on the r.h.s. is due to the fact that the laboratory frame is comoving with the crystal with the velocity V.

When the interface deforms as

$$z = \zeta(x,t), \qquad (25.7)$$

the temperature at the interface varies as

$$T_i = T_0 + G_T\zeta. \qquad (25.8)$$

Since the Hele-Shaw cell is thin, we can neglect variation of physical quantities in the directions of the thickness, y-direction. The equilibrium concentration $c_{L,eq}$ of liquid at the interface temperature T_i is determined from Eq.(25.2) with the Gibbs-Thomson curvature effect included as

$$T_i = T_M\left(1 - \frac{\tilde{\gamma}\Omega}{L}\kappa\right) - mc_{L,eq}, \qquad (25.9)$$

with L being the latent heat per volume. By assuming a local equilibrium at the interface, $c_L(x,\zeta) = c_{L,eq}$, the liquid concentration at the interface is determined from Eq.(25.8-25.9) as

$$c_L(x.\zeta) = \frac{c_\infty}{k} - \frac{\tilde{\gamma}\Omega T_M}{mL}\kappa - \frac{G_T}{m}\zeta. \qquad (25.10)$$

From the local equilibrium assumption, the concentration in the crystal $c_S(x,\zeta)$ is given by $c_S(x,\zeta) = kc_L(x,\zeta)$ with the equilibrium segregation coefficient k. On crystallization, excess mass $V_n(c_L(x,\zeta) - c_S(x,\zeta))$ is expelled from the crystal per unit time, and this excess mass has to be transported by chemical diffusion flux $-D_c\partial c/\partial n$. The mass conservation reads as

$$V_n(c_L - c_S) = -D_c\frac{\partial c}{\partial n}. \qquad (25.11)$$

We now introduce a dimensionless diffusion field in the liquid by

$$u(x,z,t) = \frac{c - c_\infty}{\Delta c} = \frac{c - c_\infty}{(k^{-1} - 1)c_\infty}, \qquad (25.12)$$

which satisfies the far field condition $u(x, z \to \infty) = 0$. For a flat interface ($\zeta = \kappa = 0$), u is 1 at the crystal-liquid boundary. The diffusion equation reduces to

$$\frac{1}{D_c}\frac{\partial u}{\partial t} = \frac{2}{l_D}\frac{\partial u}{\partial z} + \nabla^2 u \qquad (25.13)$$

with the diffusion length $l_D = 2D_c/V$. When D_c is very large, the diffusion field u relaxes to steady state $\partial u/\partial t = 0$ quickly during the slow growth of a crystal. The field u follows the steady state equation

$$\frac{2}{l_D}\frac{\partial u}{\partial z} + \nabla^2 u = 0. \tag{25.14}$$

The local equilibrium condition (25.11) is written in this dimensionless form as

$$u_{i,\mathrm{L}} = 1 - d\kappa - \frac{\zeta}{l_\mathrm{T}}. \tag{25.15}$$

A chemical capillary length

$$d = \frac{\tilde{\gamma}\Omega T_\mathrm{M}}{mL\Delta c} \tag{25.16}$$

characterizes the surface effect, and a thermal length

$$l_\mathrm{T} = \frac{m\Delta c}{G_\mathrm{T}} \tag{25.17}$$

characterizes the scale of the temperature variation. Since the concentration of the solid at the interface is

$$u_{i,\mathrm{S}} = k(u_{i,\mathrm{L}} - 1), \tag{25.18}$$

the material conservation (25.11) is written as

$$V_n\left[k + (1-k)u_{i,\mathrm{L}}\right] = -D_c\frac{\partial u}{\partial n}. \tag{25.19}$$

Here V_n is the normal growth velocity given by

$$V_n = \left(V + \frac{\partial \zeta}{\partial t}\right)n_z. \tag{25.20}$$

25.2 Stability of a Flat Interface

When the flat interface is growing steadily in z direction with a velocity V, the diffusion equation has a solution

$$u_0(z) = e^{-2z/l_D}, \tag{25.21}$$

which satisfies the boundary condition $u_0(z = \infty) = 0$ and $u_0(z = 0) = 1$. When the interface deforms to

$$z = \zeta(x,t) = a_q e^{\omega_q t}\cos qx, \tag{25.22}$$

the diffusion field also deforms as

$$u(x,z,t) = u_0(z) + A_1 e^{\omega_q t}\cos qx\, e^{-\Lambda_q z}. \tag{25.23}$$

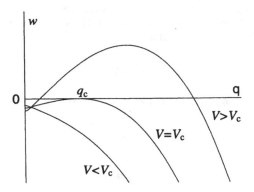

Figure 25.4: Dispersion relation of the interface modulation during the unidirectional solidification.

Here A_1 is proportional to a_q, and the damping rate in z direction, Λ_q, should satisfy the diffusion equation in quasi-stationary approximation (25.14) as

$$\frac{-2\Lambda_q}{l_D} + \Lambda_q^2 - q^2 = 0$$

or

$$\Lambda_q = l_D^{-1} + \sqrt{q^2 + l_D^{-2}}. \tag{25.24}$$

From the local equilibrium condition (25.15) A_1 is determined as

$$A_1 = \left(\frac{2}{l_D} - \frac{1}{l_T} - dq^2\right) a_q. \tag{25.25}$$

By inserting these results in the material conservation (25.19), one obtains the dispersion relation

$$\frac{\omega_q}{D_c} = \left(l_D^{-1} + \sqrt{q^2 + l_D^{-2}}\right)\left(\frac{2}{l_D} - \frac{1}{l_T} - dq^2\right) - \left(\frac{2}{l_D}\right)^2 + \frac{2}{l_D}(1-k)\left(\frac{1}{l_T} + dq^2\right), \tag{25.26}$$

which is depicted in Fig.25.4. For slow pulling velocity V, ω_q is negative and the deformation decays in time. The flat interface is stable. When the pulling rate V is larger than the critical value V_c, the maximum value of ω_q becomes positive, and the flat interface becomes unstable against those deformations with wavenumbers with $\omega_q > 0$. At the critical velocity V_c, the maximum of ω_q becomes zero: $\omega_q = \partial\omega_q/\partial q = 0$. (See Fig.25.4.) Since V_c is small, the diffusion length $l_D = 2D/V$ is large such that the relation l_D, $l_T \gg d$ holds, and the wavenumber for the most unstable mode is expected as large as $ql_D \gg 1$. Then the dispersion is approximated as

$$\frac{\omega_q}{D_c} \approx -k\left(\frac{2}{l_D}\right)^2 + q\left(\frac{2}{l_D} - \frac{1}{l_T}\right) - dq^3. \tag{25.27}$$

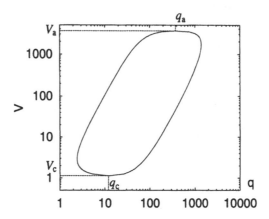

Figure 25.5: Neutral curve of the unidirectional solidification.

The critical velocity V_c is obtained as

$$\frac{V_c}{D_c} = \frac{2}{l_{D,c}} \approx \frac{1}{l_T} + \frac{3}{l_{D,c}} \left(\frac{4k^2 d}{l_{D,c}} \right)^{1/3} \approx \frac{1}{l_T} + \frac{3}{2} \left(\frac{2k^2 d}{l_T} \right)^{1/3} \frac{1}{l_T}, \tag{25.28}$$

and the critical wavenumber is

$$q_c \approx \left(\frac{2k}{dl_{D,c}^2} \right)^{1/3} \approx \left(\frac{k}{2dl_T^2} \right)^{1/3}. \tag{25.29}$$

For $V > V_c$, ω_q is positive for some regions of wavenumbers, and the flat interface is unstable for these sinusoidal modes with wavevectors within this region. The locus of $\omega_q(q, V) = 0$ represents the neutral curve $q = q_N(V)$ as shown in Fig.25.5.

The neutral curve is shown to be closed at the upper critical velocity V_a. When the pulling velocity is too fast, the diffusion length becomes of the order of the capillary length, $l_T \gg l_D \sim d$, and the interface is stabilized by the capillarity. This is called the absolute stability. For $ql_D \ll 1$, the dispersion relation is simplified as

$$\frac{\omega_q}{D_c} = -\frac{2k}{l_D l_T} + \left(1 - \frac{l_D}{2l_T} - \frac{2dk}{l_D} \right) q^2 - \frac{(d + l_D)l_D}{2} q^4. \tag{25.30}$$

At the upper critical velocity

$$\frac{V_a}{D_c} = \frac{2}{l_{D,a}} \approx \frac{1}{dk} - \sqrt{\frac{1 + 2k}{dkl_T}}, \tag{25.31}$$

the maximum value of ω_q vanishes: $\omega_q = \partial \omega_q / \partial q = 0$. For the velocity higher than V_a, ω_q always stays negative. The critical wavenumber is given by

$$q_a \propto \left[k(1 + 2k)d^3 l_T \right]^{-1/4}. \tag{25.32}$$

25.3   Eckhaus Instability

When the interface deforms sinusoidally as $z = \zeta(x,t) + \zeta^*(x,t) = \int Re(a_q e^{iqx})dq$, the Fourier transform $a_q(t)$ develops as

$$\frac{da_q}{dt} = \omega_q a_q \qquad (25.33)$$

in the linear approximation. Here ω_q is determined from the dispersion relation (25.26). When the pulling velocity V is a little larger than the critical value V_c, the interface is unstable for the sinusoidal modes with $\omega_q > 0$. Around the critical wavelength q_c, ω_q is expanded as

$$\omega_q = \omega_c - C(q - q_c)^2 \qquad (25.34)$$

with $\omega_c = \omega_{q_c}$, and $C = -\frac{1}{2}(\partial^2\omega/\partial q^2)_{q_c}$ is positive. ω_c is proportional to $V - V_c$. From Eq.(25.33) and (25.34), the time evolution of the interface $\zeta(x,t)$ can be approximated as

$$\frac{\partial\zeta(x,t)}{\partial t} = \omega_c\zeta - C\left(\frac{\partial}{i\partial x} - q_c\right)^2\zeta. \qquad (25.35)$$

Near the critical point V_c, the unstable mode has the wave numbers around q_c, and the deformation is expressed as

$$\zeta = A(x,t)e^{iq_c x} \qquad (25.36)$$

with the slowly varying complex amplitude $A(x,t)$. It satisfies the linear equation

$$\frac{\partial A}{\partial t} = \omega_c A + C\frac{\partial^2 A}{\partial x^2}. \qquad (25.37)$$

As ω_c is positive, the deformation A increases, but then the nonlinearity should come into play. Since the system is invariant by the transversal translation $x \to x + \phi$ and the space inversion $x \to -x$, one gets the nonlinear amplitude equation of the Landau-Ginzburg type,

$$\frac{\partial A}{\partial t} = \omega_c A + C\frac{\partial^2 A}{\partial x^2} - \alpha_1|A|^2 A. \qquad (25.38)$$

There is a systematic derivation of the Landau coefficient α_1 by the reductive perturbation method [199]. To specify the meaning of the Landau coefficient α_1, we consider the amplitude of the critical mode: $A(x,t) = A(t)$. If $\alpha_1 > 0$ the third order term acts as to limit the amplitude A. The amplitude increases gradually as $A \sim \omega_c^{1/2} \sim (V - V_c)^{1/2}$ from zero near the critical point, as shown in Fig.25.6. The transition is similar to the second order phase transition in equilibrium case, and is called supercritical. If $\alpha_1 < 0$, one needs still higher order terms to obtain a finite amplitude of deformation. For example, with the fifth-order term A^5 with negative

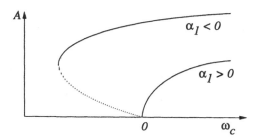

Figure 25.6: Schematics of the modulation amplitude of the interface near the instability. For a positive Landau coefficient α_1, the bifurcation from the planar to the modulated interface is supercritical, whereas for a negative α_1 it is subcritical.

coefficient, the amplitude jumps to a finite value at V_c with the hysteresis. The transition is similar to the first-order phase transition in equilibrium, and the bifurcation is called subcritical. We consider hereafter only the supercritical case with $\alpha_1 > 0$.

One can easily find that there is a stationary solution to Eq.(25.38) in the form of a single sinusoidal mode as

$$A_0(x) = A_q e^{i(q-q_c)x}. \tag{25.39}$$

Its amplitude A_q is determined as

$$A_q^2 = \frac{\omega_c - C(q - q_c)^2}{\alpha_1} = \frac{\omega_q}{\alpha_1}. \tag{25.40}$$

In the domain of $\omega_q > 0$ where the flat interface is unstable, a stationary profile of the interface is possible which is deformed sinusoidally as $z = \mathrm{Re}(A_q e^{iqx})$ with q dependent amplitude A_q. But what periodicity is selected? Are all these nonlinear solutions stable?

We consider now the linear stability of the stationary state against the modification:

$$A(x,t) = A_q e^{i(q-q_c)x} \left[1 + \xi(x,t)\right] \tag{25.41}$$

with a complex modification ξ. For small ξ, the modification can be written as $1 + \xi \sim e^{\xi} \sim e^{\mathrm{Re}\xi} e^{i\mathrm{Im}\xi}$. This means that the real part $\mathrm{Re}\xi$ represents an amplitude modification to A_q, and the imaginary part $\mathrm{Im}\xi$ represents a phase modulation. By inserting Eq.(25.41) to Eq.(25.38), and taking the first order of ξ, one gets

$$\frac{\partial \xi}{\partial t} = C \left[2i(q - q_c)\frac{\partial \xi}{\partial x} + \frac{\partial^2 \xi}{\partial x^2}\right] - \alpha_1 A_q^2(\xi + \text{c.c.}). \tag{25.42}$$

This linear equation can be analyzed by using the Fourier transformation as

$$\xi(x,t) = e^{\Omega t} \left(B_1 e^{iQx} + B_2 e^{-iQx}\right) \tag{25.43}$$

with real B_1 and B_2. By comparing coefficients for e^{iQx} and e^{-iQx}, we get the linear equation for B_1 and B_2 as

$$\Omega B_1 = C\left[-2Q(q-q_c) - Q^2\right]B_1 - \left[\omega_c - C(q-q_c)^2\right](B_1 + B_2)$$
$$\Omega B_2 = C\left[2Q(q-q_c) - Q^2\right]B_2 - \left[\omega_c - C(q-q_c)^2\right](B_1 + B_2). \quad (25.44)$$

Here $\alpha_1 A_q^2$ is replaced by the corresponding term by Eq.(25.40). On requiring that B_1 and B_2 has nontrivial solution, the eigenvalue equation

$$\begin{vmatrix} \Omega + 2CQ(q-q_c) + CQ^2 + \omega_c - C(q-q_c)^2 & \omega_c - C(q-q_c)^2 \\ \omega_c - C(q-q_c)^2 & \Omega - 2CQ(q-q_c) + CQ^2 + \omega_c - C(q-q_c)^2 \end{vmatrix}$$
$$= \Omega^2 + 2C\left[Q^2 + Q_m^2 - (q-q_c)^2\right]\Omega + 2C^2Q^2\left[Q_m^2 - 3Q^2 + \frac{1}{2}(q-q_c)^2\right]$$
$$= 0 \quad (25.45)$$

is derived. Here $Q_m^2 = \omega_c/C$. Eigenvalues are obtained as

$$\Omega_{\pm} = -C\left[Q_m^2 - (q-q_c)^2 + Q^2\right] \pm C\sqrt{[Q_m^2 - (q-q_c)^2]^2 + 4Q^2(q-q_c)^2}. \quad (25.46)$$

For a long wavelength modulation with a small wavenumber Q, one gets

$$\Omega_- \approx -2C\left[Q_m^2 - (q-q_c)^2\right] + O(Q^2) \approx -2\omega_q$$
$$\Omega_+ \approx -C\frac{Q_m^2 - 3(q-q_c)^2}{Q_m^2 - (q-q_c)^2}Q^2 \quad (25.47)$$

with corresponding eigenmodes; $B_{2,-} = B_{1,-} = B_-$ and $B_{2,+} = -B_{1,+} = B_+$, or $\xi_- = B_-e^{\Omega_- t}\cos Qx$ and $\xi_+ = iB_+e^{\Omega_+ t}\sin Qx$. Therefore, $-$ mode corresponds to the amplitude modulation and $+$ mode corresponds to the phase modulation. Since $\Omega_- < 0$ for $\omega_q > 0$, amplitude modulation always damps out. On the contrary, Ω_+ can be positive for those modes with wavenumbers q in the region

$$Q_m^2 > (q-q_c)^2 > \frac{Q_m^2}{3} = \frac{\omega_c}{3C}. \quad (25.48)$$

The phase modulation destabilizes the stationary solution. Within the region (25.48) near the neutral line $q = q_N(V)$, the phase diffusion coefficient is negative and the phase diffusion mode increases. This is called the Eckhaus instability [55]. Neutral curve $q = q_N(V)$ is determined from the condition $\omega_q = 0$. With Eq.(25.34), then, $q_{N,\pm} = q_c \pm \sqrt{\omega_c/C}$. From Eq.(25.48) the Eckhaus boundary is obtained as $q_{E,\pm} = q_c \pm \frac{1}{\sqrt{3}}(q_{N,\pm} - q_c)$, as depicted in Fig.25.7.

In the analysis the dispersion relation, Eq.(25.34), is assumed to have a maximum at the critical wavenumber q_c even in the supercritical regime $V > V_c$. In reality, the most unstable wavenumber q_m shifts rapidly depending on V. Therefore, the approximate Eckhaus boundary can be written around the most unstable mode as $q_{E,\pm} = q_m \pm \frac{1}{\sqrt{3}}(q_{N,\pm} - q_m)$. Brattkus and Misbah calculated the Eckhaus boundary more precisely, which is quite different from parabola [29].

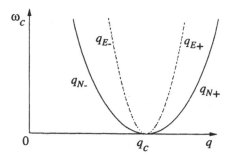

Figure 25.7: The sinusoidal modulation with wavenumbers q inside the region E is unstable against the phase diffusion, and this is called Eckhaus instability.

25.4   Fully Nonlinear Behavior

Near the critical point where the surface deformation is small, the amplitude equation (25.38) with $\alpha_1 > 0$ describes the crystal shape correctly. When the Landau coefficient α_1 is negative, or the system is far from the critical point, the deformation is no more small and the full nonlinearity has to be considered. In such a case, numerical simulation is appropriate to study the large interface deformation [183, 106, 167]. We extend the simulation algorithm explained in Appendix A24.3 to the periodically modulated surface structures under a constant temperature gradient. For a small velocity near the critical velocity V_c, the cellular structure appears as shown in Fig.25.8a. On increasing the pulling velocity, the cell groove deepens and the surface takes the form of cusp arrays, as in Fig.25.8b. Further increase of the pulling velocity sharpens the tip and the dendrite array is formed as in Fig.25.8c. Here the dendrite structure is controlled by the anisotropy of the surface tension and the tip radius and the velocity satisfies the solvability condition (24.5). The simulation results qualitatively agrees with the experimental sequences of morphology changes as shown in Fig.25.9.

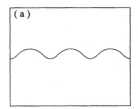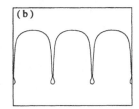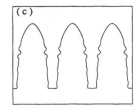

Figure 25.8: Interface structures in unidirectional solidification obtained by simulation: (a) arrays of cells ($V/V_c = 1.01$), (b) of cusps ($V/V_c = 1.15$), and (c) of dendrites ($V/V_c = 17.6$)

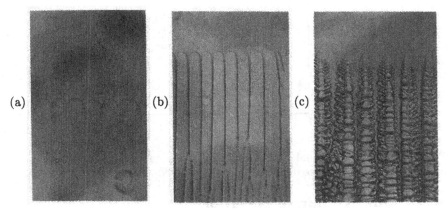

Figure 25.9: Interface structures in unidirectional solidification of succinonitrile in acetone. (a) Arrays of cells, (b) of cusps and (c) of dendrites [180].

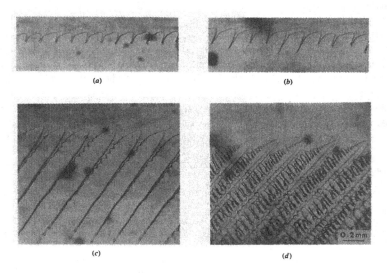

Figure 25.10: Tilted unidirectional solidification in the pivalic acid-ethanol system. Crystal axis is $\psi = 40.5°$. The velocity in $\mu m/s$ are (a) 0.5, (b) 1.0, (c) 2.75 and (d) 10.0 [182].

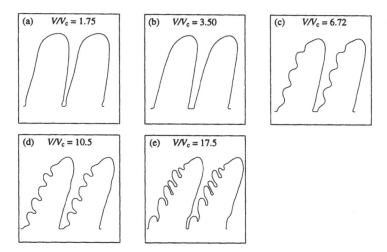

Figure 25.11: Tilted unidirectional solidification at various pulling velocities obtained by simulation. The crystal axis is $\psi = 17°$ off from the direction of the temperature gradient. The tilt angle ϕ of the tip profile is (a) 4.3°, (b) 8.7°, (c) 12.5°, (d) 14.2° and (e) 16.7° [148].

In directional solidification there are two tip stabilizing effects, the thermal gradient and the surface stiffness. What will happen when two anisotropies prefer different orientations: the temperature gradient forces the crystal to grow in the z direction, but the capillarity prefers another direction with a tilt angle ψ. It means that the capillary length depends on the orientation ψ of crystal axis, and is written as

$$d = d_0 \left[1 - \epsilon \cos 4(\theta - \psi) \right] \qquad (25.49)$$

at the point with the angle θ of the interface normal to z axis.

Tilted unidirectional solidification is observed experimentally [73, 182, 27, 2, 144]. Figure 25.10 shows interface structures at various pulling rates for the pivalic acid-ethanol system [182]. On increasing the pulling rate, the interface tilts from the direction of the temperature gradient, and approaches to that of the crystalline axis. Since the tilting increases by increasing the velocity, the surface kinetics is supposed to be relevant. Also from the linear stability analysis, the anisotropy in the surface stiffness is found to be irrelevant for the tilting [51, 202]. But the recent numerical simulations of the unidirectional solidification with local equilibrium assumption [2, 148] show that the tilting is possible when the crystalline axis is off-angled from the pulling direction. The simulation in Fig.25.11 shows that for small pulling velocity V the cell tip is oriented close to z axis, the orientation of temperature gradient, but for large V the dendritic crystal is oriented close to the crystalline axis. Thus the experimental tendency of large tilt angle at fast pulling can be reproduced by the surface stiffness, too. More studies are necessary to identify the real mechanism of crystal tilting, whether it is due to the anisotropy in capillarity or to that in the kinetics.

26   Eutectic Growth

For some alloy system such as Pb-Sn, the phase diagram looks as shown in Fig.26.1. Due to the mixing entropy the melting temperature of an alloy often decreases from the pure material. When two liquidus lines of AB alloy decrease down from the pure A and from the pure B melting temperatures by mixing the other component, they meet at some concentration of B species c_E and the temperature T_E. At T_E, the liquid solution with the concentration c_E coexists with two crystals, phase α with concentration $c_S^\alpha(< c_E)$ and phase β with concentration $c_S^\beta(> c_E)$. This triple point is called the eutectic point. At this eutectic point the melting temperature is minimum.

   If one grows this eutectic alloy unidirectionally in a Hele-Shaw cell, what kind of structure appears? Starting from the liquid with the eutectic concentration c_E, crystallization of α or β phase alone cannot satisfy the material conservation. α and β phases should appear simultaneously. It may intuitively expected that α and β phases appear alternatively as lamellae parallel to the growth direction: When the crystal is growing in positive z directions, the symmetry in x and $-x$ directions yields that the $\alpha\beta$ phase boundary is expected to align in z direction and perpendicular to the liquid-crystal interface in x direction. This is the structure found in experiment [91] at low pulling velocity, as shown in Fig.26.2 [108]. The problem to be posed is the selection of the periodicity λ.   There is a detailed calculation by Jackson and

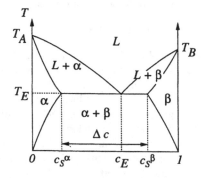

Figure 26.1: Equilibrium phase diagram of an eutectic AB alloy.

Figure 26.2: Lamellar growth of eutectic crystal, CBr$_4$- C$_2$Cl$_6$ [108]

Hunt, which is reproduced in Appendix A26 [91]. Here a phenomenological discussion is given on the periodicity selection.

26.1 Lamellar Structure

Since the thermal diffusivity is larger than the chemical diffusivity, the temperature gradient $G_T = dT/dz$ can be assumed constant. By choosing the origin of z axis at the point where the temperature is at the eutectic value T_E, the temperature distribution is described as

$$T(z) = T_E + G_T z. \qquad (26.1)$$

The liquid ($z > 0$) is hot and the crystal ($z < 0$) is cold. We consider the simplest case that a crystal is growing from a solution with eutectic concentration c_E. When the crystal takes a lamellar structure with a periodicity λ, the ratio of α phase η and that of β phase $1 - \eta$ should satisfy the conservation relation

$$c_E = \eta c_S^\alpha + (1 - \eta)c_S^\beta \qquad (26.2)$$

or

$$\eta = \frac{c_S^\beta - c_E}{\Delta c} \qquad (26.3)$$

with the miscibility gap

$$\Delta c = c_S^\beta - c_S^\alpha. \qquad (26.4)$$

Since c_S^α is smaller than c_E, B atoms are expelled out from the α crystal into the liquid by crystallization, and the liquid concentration c_L^α in front of α phase increases above c_E. On the contrary, in front of the β phase, B atoms are sucked up by β crystal and the liquid concentration c_L^β decreases below c_E during crystallization. As is apparent from the magnified phase diagram Fig.26.3 with metastable liquid branches, deviation of the liquid concentration from the eutectic value means that the supercooling at the interface under the local equilibrium assumption. The crystal concentration c_S^α and

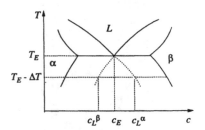

Figure 26.3: Phase diagram near the eutectic temperature T_E with metastable branches of coexistence curve.

c_S^β correspondingly differ from those at the eutectic temperature, but their differences are small and do not modify the conclusion essentially.

Since the concentration in front of each crystal is different from that at $z \to \infty$ or c_E, the concentration diffusion takes place. The diffusion flow compensates the material deficit or surplus produced by the crystallization. In front of the α crystal, $v_n(c_L^\alpha - c_S^\alpha) \approx v_n(c_E - c_S^\alpha)$ of B atoms are expelled from the unit surface per unit time. This excess is transported by the diffusion flow $-D_c \partial c/\partial n$. However, the material does not need to be transported up to $z \to \infty$: It is sufficient to be transported only to the neighboring lamellae of β phase, where the material is deficient. Since the concentration difference in liquid in front of the α and β phases is $c_L^\alpha - c_L^\beta$ for a separation of $\lambda/2$, the growth rate is determined as

$$(c_E - c_S^\alpha)v \approx Q_\alpha D_c \frac{c_L^\alpha - c_L^\beta}{\lambda/2}. \tag{26.5}$$

If the interface is flat, the direction of material diffusion x is orthogonal to the growth direction z and the diffusion does not contribute to the growth. Actually, the interface is curved, and the contribution to growth appears. This is taken into consideration in Eq.(26.5) by the factor Q_α. Its detailed form is explained in Appendix A26. In consideration of the material conservation in front of the β phase, we get the similar formula with coefficient Q_β. In the steady state where the growth velocity of α and β phases are the same, the coefficients should satisfy the relation $(1 - \eta)/Q_\alpha = \eta/Q_\beta$ due to the conservation (26.2).

The liquid-crystal interface is curved in order to satisfy the mechanical force balance at the triple point, where three phases, liquid, α crystal, and β crystal phases, meet (Fig.26.4). The force balance in x and z directions is described as:

$$\gamma_{L\alpha} \sin\theta_\alpha + \gamma_{L\beta} \sin\theta_\beta = \gamma_{\alpha\beta} \tag{26.6a}$$

$$\gamma_{L\alpha} \cos\theta_\alpha = \gamma_{L\beta} \cos\theta_\beta. \tag{26.6b}$$

Here $\gamma_{L\alpha}$ is the surface tension between the liquid and the α phases which is assumed isotropic for simplicity, and so on. In equilibrium the liquid–α (β) crystal interface is

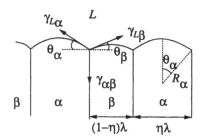

Figure 26.4: Interface profile near the triple point for the eutectic growth.

part of a circle with a radius R_α (R_β), and the angle θ_α (θ_β) is related to R_α (R_β) and the periodicity λ as

$$2R_\alpha \sin \theta_\alpha = \eta\lambda, \qquad 2R_\beta \sin \theta_\beta = (1-\eta)\lambda, \qquad (26.7)$$

as is evident from Fig.26.4.

The interface temperature is given by the concentration deviation from the eutectic values and also by the Gibbs-Thomson curvature effect as

$$T^\alpha = T_E - m_\alpha(c_L^\alpha - c_E) - \frac{\gamma_{L\alpha}T_E}{L_\alpha}\frac{1}{R_\alpha} \qquad (26.8a)$$

$$T^\beta = T_E + m_\beta(c_L^\beta - c_E) - \frac{\gamma_{L\beta}T_E}{L_\beta}\frac{1}{R_\beta}. \qquad (26.8b)$$

Here $m_\alpha = |dT_\alpha/dc_L^\alpha|$ and $m_\beta = |dT_\beta/dc_L^\beta|$ are the absolute slopes of the liquidus lines for α and β phases, L_α, L_β are latent heats of α and β phases, respectively.

By adding Eq.(26.8a) and (26.8b), liquid concentration difference is written as

$$c_L^\alpha - c_L^\beta = \left(\frac{\Delta T^\alpha}{m_\alpha} + \frac{\Delta T^\beta}{m_\beta}\right) - \left(\frac{\gamma_{L\alpha}T_E}{m_\alpha L_\alpha}\frac{1}{R_\alpha} + \frac{\gamma_{L\beta}T_E}{m_\beta L_\beta}\frac{1}{R_\beta}\right). \qquad (26.9)$$

Here the interface supercoolings are defined as $\Delta T^\alpha = T_E - T^\alpha$ and $\Delta T^\beta = T_E - T^\beta$.

We consider the simplest case where crystal phases α and β are symmetric: $c_E = 1/2$, $m_\alpha = m_\beta = m$, $\theta_\alpha = \theta_\beta = \theta$, $\eta = 1/2$, $Q_\alpha = Q_\beta = Q$, $\gamma_{L\alpha} = \gamma_{L\beta} = \gamma$, $\Delta T^\alpha = \Delta T^\beta = \Delta T$, and $c_L^\alpha - c_S^\alpha = \Delta c/2$. By using the capillary length $d = \gamma_{L\alpha}T_E/Lm\Delta c$, the interface supercooling is written from Eqs.(26.5) and (26.9) as

$$\Delta T = T_E\left(a_D\frac{\lambda}{l_D} + a_K\frac{d}{\lambda}\right), \qquad (26.10)$$

which consists of two contributions; Diffusional one $\Delta T_D = T_E a_D \lambda/l_D$ and the kinetic one $\Delta T_K = T_E a_K d/\lambda$. Here $l_D = 2D_c/v$ is the diffusion length, and material parameters are represented as $a_D = m\Delta c/4QT_E$, $a_K = 4\sin\theta m\Delta c/T_E$. The relation between ΔT and λ is shown in Fig.26.5.

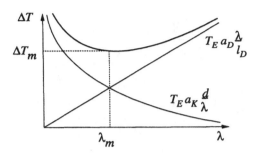

Figure 26.5: Minimum undercooling hypothesis for the selection of the eutectic lamellar periodicity.

In order to determine the lamellar periodicity λ, the minimum supercooling hypothesis is used by Jackson and Hunt: $\partial \Delta T / \partial \lambda = 0$. Then the periodicity is determined as

$$\lambda_m = \sqrt{\frac{a_K}{a_D}} \sqrt{l_D d} \approx \frac{1}{\sqrt{16 Q \sin \theta}} \sqrt{l_D d}. \qquad (26.11)$$

This yields the relation between the pulling velocity v and the periodicity λ_m as

$$v \lambda_m^2 = \frac{D_c d}{8 Q \sin \theta} = const. \qquad (26.12)$$

Also the interface undercooling ΔT_m is determined as

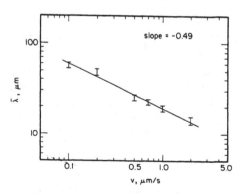

Figure 26.6: Variation of the average interlamellar spacing $\bar{\lambda}$ with the growth velocity v for the carbon tetrabromide (CBr$_4$)- hexachloroethene (C$_2$Cl$_6$) eutectic system [174].

$$\Delta T_m = 2T_{\rm E}\sqrt{a_{\rm D}a_{\rm K}}\sqrt{\frac{d}{l_{\rm D}}} \tag{26.13}$$

and it satisfies the relation with the pulling velocity v as

$$\frac{\Delta T_m^2}{v} = \frac{2\sin\theta}{Q}(m\Delta c)^2\frac{d}{D_{\rm c}} = const. \tag{26.14}$$

The full nonlinear analysis by computer simulation [107, 95, 98, 96] also show that the interface undercooling has the minimum as a function of the periodicity, and that the Jackson and Hunt theory provides the accurate position of the minimum undercooling [96]. The scaling relations, (26.12) and (26.14), are also shown there [98, 96]. These relations are also confirmed by experiments, as shown in Fig.26.6 [174, 203].

26.2 Parity Breaking and Oscillation

On increasing the pulling velocity V or on varying the liquid concentration from the eutectic value, the lamellar structure is found to be modified.

For slow pulling rate, the phase boundary between α and β phases align parallel to the temperature gradient. On suddenly increasing the pulling rate by a factor 4, the periodicity selected by the minimum undercooling condition should be half of the initial one, as Eq.(26.12) tells. But such a large structural variation is not possible topologically. The numerical simulation using the boundary element method in the stationary code found that the parity (left-right symmetry) will break for large pulling rate [52, 95, 97]: The $\alpha\beta$ or $\beta\alpha$ triple point shifts transversally along the liquid-crystal interface, and the $\alpha\beta$ phase boundary tilt from the temperature gradient, as shown in Fig.26.7a. The tilting is found later in many experiments (Fig.26.7b) [57, 58].

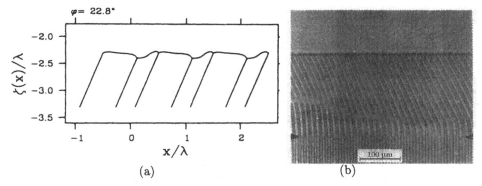

(a)   (b)

Figure 26.7: Parity broken lamellae in eutectics obtained (a) by simulation and [97] (b) by experiment of $CBr_4-C_2Cl_6$ [57].

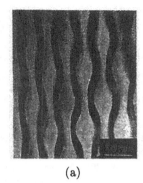
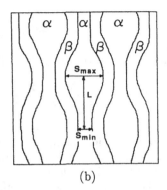

(a)   (b)

Figure 26.8: Periodic oscillation of lamellae in off-eutectic alloy (a) by the experiment of Al-CuAl$_2$, and (b) by simulation [204].

This tilting is different from that discussed in the previous section 25.3. There the tilting is enforced by the tilting of crystalline axis from the growth direction and the parity is externally broken. Here the system has the left-right parity symmetry, but the realized state breaks the parity spontaneously.

By shifting the far field concentration c_∞ from the eutectic value c_E, system shows the nonstationary structure, as periodic oscillation shown in Fig.26.8a [204]. The linear stability analysis showed the instability of the oscillatory motion of the triple point [53, 94]. There is a model simulation which shows the oscillatory behavior, and the result is in good agreement to the experiment as shown in Fig.26.8b [204].

27   Diffusion Effect on Polyhedral Crystal: Berg Effect

So far it is assumed that the surface kinetics is infinitely fast and the local equilibrium is realized at the interface in order to stress the diffusion effect on the front instability. This approximation is not far from reality for an ice crystal growing in water [65, 66]. The prism face of ice in water is rough, and the heat transport controls the growth. On the other hand, for snow growing from the water vapor, the approximation does not hold since the surface of the snow is sharp and faceted. The kinetics is slow and has to be properly taken into account.

If the interface is atomically rough, the kinetic coefficient K in Eq.(18.8) is finite but not singular. The growth shape can be anisotropic but is nonsingular and smooth. The orientation dependence of K leads to the new scaling relation of the dendritic growth rate and the tip radius [32, 33, 50, 169]. When the preferred orientations of the surface tension and of the kinetic coefficient are different, rich variety of growth shapes with dynamical shape transitions is expected [45]

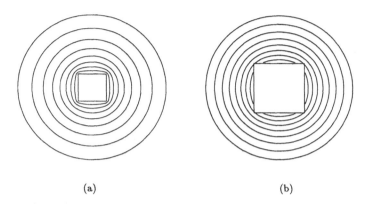

(a)   (b)

Figure 27.1: (a)Electrostatic potential around metallic polygon. The metal surface is equipotential. (b) Concentration field distribution around the square crystal growing steadily with a constant normal gradient $q_n = \partial c/\partial n$ along the surface.

If the interface is atomically flat, the kinetic coefficient K is singular, and the crystal is expected to take polyhedral morphology covered with flat facets of singular surfaces. The growth of these singular surfaces is governed by the surface kinetics discussed in Part III. However, material has to be transported to the growing interface by the diffusion, and the diffusion also influences the growth form. One has to consider both the kinetic and diffusion processes in this situation.

First we consider the concentration distribution around the polyhedral crystal growing from the solution. It is determined by Eq.(18.6-18.8). Usually, crystal growth rate is small for kinetic-controlled growth, and one can assume the stationary condition that the diffusion field quickly adjust its distribution around the given crystal morphology: $\partial u/\partial t = 0$. Thus the diffusion field u satisfies the Laplace equation

$$\nabla^2 u = 0 \qquad (27.1)$$

instead of the diffusion equation (18.6). An analogy holds between the concentration distribution $u(\mathbf{r})$ and the electrostatic potential $\phi(\mathbf{r})$. Far from the crystal, the concentration distribution reflects the isotropic nature of the system and equi-concentrations are asymptotically spherical. If the concentration is assumed constant on the polyhedral face, $u(\mathbf{r})$ corresponds to electrostatic potential $\phi(\mathbf{r})$ around the metallic polyhedron, and looks as shown in Fig.27.1a. Near the corner of the polyhedron, the spacing between consecutive equi-concentrations are narrow, and the normal growth rate determined by Eq.(18.7) or $V_n = -D_c \partial c/\partial n$ is larger at the corner than at the center. Then the polyhedral face cannot remain flat. To keep the polyhedral

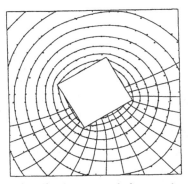

Figure 27.2: Concentration distribution around the growing NaCl crystal from water solution [24].

face flat, there should be concentration difference on the surface, especially between those at the corner and at the center.

If we assume the constant normal gradient on a flat but bounded surface, the concentration at the corner should be higher than that at the center, as shown in Fig.27.1b. This was shown analytically in two dimensions [173]. Experimentally, the concentration profile around the polyhedral crystal is observed by Berg as shown in Fig.27.2, and the concentration difference along the surface is observed. This effect is called the Berg effect [24]. The distribution of the supersaturation along the flat surface is also measured [42].

If the face is really flat and the concentration varies on it, the kinetically controlled growth rate (18.8) varies from position to position, and it is impossible to keep the face flat. To realize a steady growth, the kinetic coefficient should vary along the surface. This is possible when a macroscopically flat surface contains many microscopic steps and is in fact a vicinal face. Then the kinetic coefficient depends on the concentration. For example, at the corner of the crystal where the supersaturation is high, the two-

Figure 27.3: Formation of snow dendrite by simulation [201].

dimensional nucleation supplies steps toward the center of the face [119]. The step advancement velocity v decreases near the center where the supersaturation is low, but if the step density is high and the step separation ℓ is small, the normal growth rate $R = av/\ell$ can be kept constant over the surface.

If the supersaturation at the corner is too high or the crystal has grown too large, the difference in supersaturations at the corner and at the center becomes too large to keep the constant normal growth rate over the flat face. The corner starts to grow faster than the center, and the instability sets in [47, 119]. This initiation of the corner instability is analyzed numerically, and the corner instability and its size dependence is found, if the nucleation controls the kinetics [119]. Also by using the boundary element method the growth of a kinetically-controlled two-dimensional crystal is simulated. The circular crystal becomes polygonal, and then produces the faceted arm from the corner of polygon, similar to the snow growth, as shown in Fig.27.3 [201]. For the sidebranch formation of a faceted dendrite, however, more studies seem necessary.

Part V
Appendices

A9 More on Surface Roughening

A9.1 Solid-on-solid (SOS) model

In order to focus on the temperature dependence of the surface structure, the solid-on-solid (SOS) model is often used [112, 43]. The model picks up the freedom of surface height $z(i)$ such that no vacancy in the crystal, no overhangs at the surface and no crystal clusters in the vapor are allowed. Here i is the site in a d-dimensional hypercubic lattice with $2d$ nearest neighbors, and the crystal occupies $d+1$ dimensional space. In a unit of an atomic size a, the height $h(i) = z(i)/a$ takes integer values, $h(i) = -\infty, \cdots, -1, 0, 1 \cdots, \infty$. Since the height difference between neighboring sites costs energy, the Hamiltonian $\mathcal{H}$ is written as

$$\mathcal{H} = J \sum_{<ij>} |h(i) - h(j)|^p. \tag{A9.1}$$

Here the model with $p = 1$ is called the (absolute) SOS model, the model with $p = 2$ the discrete Gaussian (DG) model. The latter is convenient to treat analytically. If the height takes only two values, $h(i) = 0, +1$, the model corresponds to the Ising model.

As to measure the surface roughness, the height difference correlation function [49]

$$G(r_{ij}) = \langle (h(i) - h(j))^2 \rangle \tag{A9.2}$$

is used. When the surface is flat, heights are strongly correlated and G remains finite even for $r_{ij} \to \infty$. When the surface is rough, heights at different positions fluctuate independently, and $G(r \to \infty)$ diverges. In an Ising model, G can at most be 1, and the surface can never be rough in this sense.

A9.2 Monte Carlo simulation of the SOS model

This appendix contains a source list of an example program of Monte Carlo simulation explained in Section 9.1. It is hoped that it provides a first step for the reader to get familiar with the simulation and to develop one's own program. Random number generator is copied from the reference [67].

```
      program sos
c===============================================
c    Monte Carlo simulation of 2D-sos model for roughening
c===============================================
      parameter (lxmx=128,lymx=128)
```

```
      dimension ih(lxmx,lymx),prob(6)
      read(*,*) lx,ly,loop,temp,field,iseed
c------------------------------------------------------------------
c  ih(ix,iy);      solid height
c     lxmx,lymx;  maximum linear size in x- and y-direction
c     lx,ly     ; actual linear size of the system
c  prob(k);   transition probability
c             k=1  for crystallization
c             k=2-6 for desorption with k-2 neighbours
c  temp;  temperature, (coupling J=1)
c field;  chemical potential difference
c  loop;  loop number for small accumulation and averaging
c iseed;  seed of the random number
c   mag;  magnitude of height sum
c  nnex;  nearest neighbor energy
c-------------------------------------------
      call ranint(iseed)
      call init(lxmx,lymx,lx,ly,ih)
      write(*,*) 'lx,ly,loop,temp,field;'
      write(*,*) lx,ly,loop,temp,field
      write(*,*) 'ilp,mag,nnex'
      call engy(lxmx,lymx,lx,ly,ih,mag,nnex)
      call prbini(prob,temp,field)
c-------------- monte carlo loop
      do 10 ilp=1,loop
      do 11 inlp=1,lx*ly
      call mc(lxmx,lymx,lx,ly,ih,prob,mag,nnex)
   11 continue
      write(*,*)\ilp,mag,nnex
   10 continue
c----------- check the final height and exchange energy
      call engy(lxmx,lymx,lx,ly,ih,mag1,nnex1)
      write(*,*) mag,mag1,nnex,nnex1
      stop
      end
c===================================================
      subroutine init(lxmx,lymx,lx,ly,ih)
c------------------------------------------------------------------
c      initiallization
c------------------------------------------------------------
      dimension ih(lxmx,lymx)
      do 200 iy=1,ly
      do 200 ix=1,lx
```

```
      ih(ix,iy)=0
  200 continue
      return
      end
c==================================================
      subroutine mc(lxmx,lymx,lx,ly,ih,prob,mag,nnex)
c--------------------------------------------------------
c      Monte Carlo step
c--------------------------------------------------------
      dimension ih(lxmx,lymx),prob(6)
      isum=0
      ix=randm()*lx+1
      iy=randm()*ly+1
      ixp=ix+1
      ixm=ix-1
      if(ixp.gt.lx) ixp=ixp-lx
      if(ixm.lt.1)  ixm=ixm+lx
      iyp=iy+1
      iym=iy-1
      if(iyp.gt.ly) iyp=iyp-ly
      if(iym.lt.1)  iym=iym+ly
      if(ih(ixp,iy).lt.ih(ix,iy)) isum=isum+1
      if(ih(ix,iyp).lt.ih(ix,iy)) isum=isum+1
      if(ih(ixm,iy).lt.ih(ix,iy)) isum=isum+1
      if(ih(ix,iym).lt.ih(ix,iy)) isum=isum+1
      trpr=prob(1)+prob(isum+2)
      trtr=randm()
c--------------- adsorption
      if(trtr.lt.prob(1)) then
      ih(ix,iy)=ih(ix,iy)+1
      jbup=0
      if(ih(ixp,iy).lt.ih(ix,iy)) jbup=jbup+1
      if(ih(ix,iyp).lt.ih(ix,iy)) jbup=jbup+1
      if(ih(ixm,iy).lt.ih(ix,iy)) jbup=jbup+1
      if(ih(ix,iym).lt.ih(ix,iy)) jbup=jbup+1
      mag=mag+1
      nnex=nnex+2*jbup-4
      return
      endif
c--------------- desorption
      if(trtr.lt.trpr) then
      ih(ix,iy)=ih(ix,iy)-1
      mag=mag-1
```

```
      nnex=nnex+4-2*isum
      return
      endif
c-------------- otherwise
      return
      end
c=======================================================
      subroutine engy(lxmx,lymx,lx,ly,ih,mag,nnex)
c-------------------------------------------------------
c        energy calculation
c-------------------------------------------------------
      dimension ih(lxmx,lymx)
      mag=0
      nnex=0
      do 100 ix=1,lx
      ixp=ix+1
      if(ixp.gt.lx) ixp=ixp-lx
      do 100 iy=1,ly
      iyp=iy+1
      if(iyp.gt.ly) iyp=iyp-ly
      mag=mag+ih(ix,iy)
      nnex=nnex+iabs(ih(ix,iy)-ih(ix,iyp))+iabs(ih(ix,iy)-ih(ixp,iy))
  100 continue
      return
      end
c=======================================================
      subroutine prbini(prob,temp,field)
      dimension prob(6)
      ef=exp(field/temp)+exp(4./temp)
      prob(1)=exp(field/temp)/ef
      do 10 nn=2,6
      prob(nn)=exp(2.*(nn-4.)/temp)/ef
   10 continue
      return
      end
c=======================================================
      subroutine ranint(ix)
      common /rand/ m,j
      dimension m(521),ia(521)
      do 10 i=1,521
      ix=69069*ix
   10 ia(i)=sign(1,ix)
      do 20 j=1,521
```

```
      ip=mod((j-1)*32,521)+1
      m(j)=0
      do 30 i=1,31
      ii=mod(ip+i-2,521)+1
      m(j)=2*m(j)+(ia(ii)-1)/(-2)
      ij=mod(ii+488,521)+1
  30  ia(ii)=ia(ii)*ia(ij)
      ii=mod(ip+30,521)+1
      ij=mod(ii+488,521)+1
  20  ia(ii)=ia(ii)*ia(ij)
      j=0
      return
      end
c=================================
      function randm()
      common /rand/ m,j
      dimension m(521)
      ip=521
      iq=489
      j=j+1
      if(j.gt.ip) j=1
      k=j+iq
      if(k.gt.ip) k=k-ip
      m(j)=ieor(m(k),m(j))
      randm=m(j)*0.4656612e-9
      return
      end
```

A9.3 Continuous Gaussian model

If the height variable $h(i)$ does not take the discrete integer values but takes continuous values, and the energy is given by (A9.1) with $p = 2$, the model is called the continuous Gaussian model. The partition function and correlation function of the model can be calculated straightforwardly.

The partition function is defined as

$$Z = \Pi_{i=1}^{N} \int_{-\infty}^{\infty} dh(i) \exp\left[-J \sum_{<ij>} \frac{(h(i) - h(j))^2}{k_B T}\right]. \tag{A9.3}$$

One can introduce the Fourier transformation as

$$h(i) = \frac{1}{\sqrt{N}} \sum_q e^{-iqr_i} h(q) \tag{A9.4}$$

with wavenumbers $q = 2\pi n/L$ and $n = -(L-1)/2, \cdots, 0, 1, \cdots, L/2$. Here L is the linear dimension of the system such that $N = L^d$. The coefficient $h(q)$ is given by

$$h(q) = \frac{1}{\sqrt{N}} \sum_i e^{iqr_i} h(i). \tag{A9.5}$$

In terms of $h(q)$, the partition function is written as

$$Z = \Pi_q \int_{-\infty}^{\infty} dh(q) \exp \left[-\frac{1}{2} \sum_q \frac{|h(q)|^2}{G_0(q)} \right] = \Pi_q \sqrt{2\pi G_0(q)}. \tag{A9.6}$$

Here $G_0(q)$ is the lattice Green's function defined by

$$G_0(q) \equiv \frac{k_B T}{2J} \frac{1}{\sum_\delta (1 - e^{i\delta q})} \approx \frac{k_B T}{2J} \frac{1}{q^2}, \tag{A9.7}$$

where δ is a vector connecting nearest neighbor sites. Surface free energy is obtained as

$$F = -\frac{1}{2} k_B T \sum_q \ln \left[2\pi G_0(q) \right] = -\frac{N}{2} k_B T \int \frac{d^d q}{(2\pi)^d} \ln 2\pi G_0(q). \tag{A9.8}$$

In the last equality, the summation is replaced by the integration for a large system size N as $\sum_q = N \int_{-\pi}^{\pi} d^d q/(2\pi)^d$. The height difference correlation function is calculated for large r as [115]

$$G(r) = \langle [h(r) - h(0)]^2 \rangle = \frac{2}{N} \sum_q \langle |h(q)|^2 \rangle (1 - \cos qr) = 2 \int_{-\pi}^{\pi} \frac{d^d q}{(2\pi)^d} G_0(q)(1 - \cos qr)$$

$$\sim \begin{cases} \dfrac{k_B T}{J} \dfrac{1}{2^{d-1} \pi^{d/2} \Gamma(d/2)} \dfrac{r^{2-d} - 1}{2 - d} + const + O(r^{1-d}) & \text{for } d \neq 2 \\ \dfrac{k_B T}{2\pi J} \ln r + const & \text{for } d = 2 \,, \end{cases} \tag{A9.9}$$

where $\Gamma(d/2)$ is the Gamma function [1]. The d-dimensional surface with $d \leq 2$ is always rough because the height difference correlation function diverges for two separate points, $r \to \infty$. On the contrary, for $d > 2$ the surface is always smooth.

A9.4 Variation approximation

Continuous Gauss model shows that the surface is rough in two dimensions. In the real crystal with $d = 2$, the surface can be both smooth and rough. The discrepancy is due to the neglect of the discreteness in the surface position in the continuous Gaussian model. The height should be quantized in unit of atomic size a. For analytical treatment on the phase transition and critical phenomena, a modification is introduced in the model Hamiltonian as

$$\mathcal{H} = J \sum_{<ij>} [h(i) - h(j)]^2 - U \sum_i \cos (2\pi h(i)), \tag{A9.10}$$

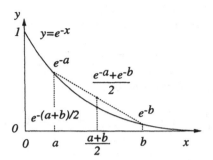

Figure A9.1: Convexity of exponential function.

where the height variables $h(i)$ are continuous, but the additional one-body potential with a small positive coefficient U prefers the height to take integer values.

We here study the phase transition of the model (A9.10) by means of the variation method [59]. The theory depends on the convexity of the exponential function; $e^{-(a+b)/2} \leq (e^{-a}+e^{-b})/2$. This inequality is obvious from Fig.A9.1. When the stochastic variable x is distributed according to the probability $\Pr(x)$, the average $\langle x \rangle$ is given by $\langle x \rangle = \int \Pr(x)x\,dx$, and the inequality is generalized to $e^{-\langle x \rangle} \leq \langle e^{-x} \rangle$.

For a thermodynamic system with a Hamiltonian $\mathcal{H}_0$, the canonical probability is defined by

$$\Pr = Z_0^{-1} \exp\left(-\frac{\mathcal{H}_0}{k_B T}\right) \tag{A9.11}$$

with

$$Z_0 = \mathrm{Tr}\exp\left(-\frac{\mathcal{H}_0}{k_B T}\right) = \exp\left(-\frac{F_0}{k_B T}\right). \tag{A9.12}$$

When the true system is described by the Hamiltonian

$$\mathcal{H} = \mathcal{H}_0 + V, \tag{A9.13}$$

one can show that

$$\exp\left(-\frac{\langle V \rangle_0}{k_B T}\right) \leq \langle \exp\left(-\frac{V}{k_B T}\right) \rangle_0 = \exp\left(\frac{F_0}{k_B T}\right) \mathrm{Tr}\exp\left(-\frac{\mathcal{H}_0+V}{k_B T}\right)$$

$$= \exp\left(\frac{F_0}{k_B T}\right)\exp\left(-\frac{F}{k_B T}\right). \tag{A9.14}$$

Therefore, the true free energy F satisfies the following inequality

$$F \leq F_0 + \langle V \rangle_0 \equiv F^*. \tag{A9.15}$$

If we make an appropriate choice of $\mathcal{H}_0$ which minimizes F^*, then it should be a good approximation to the true free energy F. In order to make the configuration summation tractable, we choose the effective Hamiltonian $\mathcal{H}_0$ as

$$\mathcal{H}_0 = \frac{1}{2} k_B T \sum_q \frac{|h(q)|^2}{G(q)}, \tag{A9.16}$$

where q−dependent parameters $G(q)$ are to be determined to find a minimum F^*. With this choice of $\mathcal{H}_0$, the probability Pr given by Eq.(A9.11) is Gaussian with $\langle h(q) \rangle = 0$ and $\langle |h(q)|^2 \rangle = G(q)$.

F_0 is calculated straightforwardly and is equal to (A9.8) with the replacement of $G_0(q)$ by $G(q)$. $\langle \mathcal{H}_0 \rangle_0$ is simply $N k_B T/2$. In $\langle \mathcal{H} \rangle_0$, the average of the first term is easily calculated as

$$\frac{J}{k_B T} \sum_{<ij>} \langle (h(i) - h(j))^2 \rangle = \frac{J}{k_B T} \sum_q \sum_\delta \langle |h(q)|^2 \rangle (1 - e^{-iq\delta}) = \frac{1}{2} \sum_q \frac{G(q)}{G_0(q)}. \tag{A9.17}$$

Since there are only first and second order cumulants for a Gaussian distribution, the term containing the one-body potential can be calculated by using the relation

$$\langle \cos 2\pi h \rangle_0 = \mathrm{Re} \langle e^{2\pi i h} \rangle_0 = \exp \left[i2\pi \langle h \rangle_0 - \frac{(2\pi)^2}{2} \langle h^2 \rangle_0 \right] = \exp \left[-2\pi^2 \langle h^2 \rangle_0 \right] \tag{A9.18}$$

with

$$\langle h^2 \rangle_0 = \frac{1}{N} \sum_q \langle |h_q|^2 \rangle_0 = \frac{1}{N} \sum_q G(q). \tag{A9.19}$$

The approximate free energy F^* is obtained as

$$\begin{aligned}
\frac{F^*}{k_B T} &= \frac{F_0 + \langle (\mathcal{H} - \mathcal{H}_0) \rangle_0}{k_B T} \\
&= -\frac{1}{2} \sum_q \ln \left[2\pi G(q) \right] + \frac{1}{2} \sum_q \frac{G(q)}{G_0(q)} - \frac{NU}{k_B T} \exp \left[-\frac{2\pi^2}{N} \sum_q G(q) \right] - \frac{N}{2}.
\end{aligned} \tag{A9.20}$$

If one calculates the higher order term $\langle V^n \rangle_0$, it is proportional to U^n, and the present approximation is good for a small perturbation U. To minimize, F^* is differentiated by $G(q)$ as

$$\frac{\partial (F^*/k_B T)}{\partial G(q)} = -\frac{1}{2} \frac{1}{G(q)} + \frac{1}{2} \frac{1}{G_0(q)} + \frac{2\pi^2 U}{k_B T} \exp \left[-\frac{2\pi^2}{N} \sum_q G(q) \right] = 0. \tag{A9.21}$$

Since the last term is independent of the wave number q, the q-dependent coupling $G(q)$ has the form

$$G(q) = \frac{1}{G_0(q)^{-1} + K^{-1} \xi^{-2}} \approx \frac{K}{q^2 + \xi^{-2}} \tag{A9.22}$$

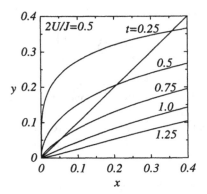

Figure A9.2: Determination of the correlation length ξ or the parameter $x = (\pi/\xi)^2$ at various temperatures $t = \pi k_B T/4J$.

where $K^{-1} = 2J/k_B T$, and the correlation length ξ is determined as

$$
\begin{aligned}
K^{-1}\xi^{-2} &= \frac{(2\pi)^2 U}{k_B T} \exp\left[-2\pi^2 \int_{-\pi}^{\pi} \frac{d^2 q}{(2\pi)^2} \frac{K}{q^2 + \xi^{-2}}\right] \\
&\approx \frac{4\pi^2 U}{k_B T} \left(\frac{\xi^{-2}}{\pi^2 + \xi^{-2}}\right)^{\pi K/2}.
\end{aligned} \tag{A9.23}
$$

By denoting $x = (\pi\xi)^{-2}$, $t = \pi K/2 = \pi k_B T/4J$, Eq.(A9.23) is transformed as

$$
x = \frac{2U}{J} \left(\frac{x}{1+x}\right)^t \equiv f(x). \tag{A9.24}
$$

Two curves $y = x$ and $y = f(x)$ are depicted in Fig.A9.2 at various temperatures t. The intersection of these two curves gives the solution of Eq.(A9.24). The slope of $f(x)$ at $x = 0$ changes drastically at $t = 1$, since

$$
\frac{df(x)}{dx} = \frac{2U}{J} t \left(\frac{1}{x} - \frac{1}{1+x}\right) \left(\frac{x}{1+x}\right)^t \sim x^{t-1}. \tag{A9.25}
$$

For small U and $t > 1$ there is only a trivial solution $x = 0$, but for $t < 1$ there is another nontrivial solution near $x \sim 0$. For small x, the solution is approximated as

$$
x = \left[\frac{J}{2U}(1+x)^t\right]^{1/(t-1)} \sim \left(\frac{J}{2U}\right)^{-1/(1-t)} \sim \exp\left[-\frac{\ln(J/2U)}{1-t}\right]. \tag{A9.26}
$$

For $t < 1$ the free energy F^* for $x \neq 0$ is smaller than that for $x = 0$, and the state with a finite ξ is stabler than the state with $\xi = \infty$. Therefore, the phase transition takes place at $t = 1$ or

$$
k_B T_R = \frac{4}{\pi} J. \tag{A9.27}
$$

The height difference correlation function is calculated by

$$G(r) = 2 \int \frac{d^2q}{(2\pi)^2} G(q)(1 - e^{iqr}) \approx \frac{K}{\pi} \int_{\min(\pi/r, \pi/\xi)}^{\pi} \frac{q \, dq}{q^2 + \xi^{-2}}$$

$$\approx \frac{k_B T}{2\pi J} \ln \left[\min(r, \xi) \right]. \tag{A9.28}$$

Above the roughening temperature $T_R \equiv 4J/\pi k_B$ or $t > 1$, there is only a solution $x = 0$ (or $\xi = \infty$), and the height fluctuation diverges at $r \to \infty$: The surface is rough. On the other hand, for $T < T_R$ or $t < 1$, ξ is finite and the height correlation function saturates as

$$G(r \gg \xi) \sim \frac{k_B T}{2\pi J} \ln \xi \sim (1 - t)^{-1} \sim \left(1 - \frac{T}{T_R} \right)^{-1} \tag{A9.29}$$

for $r \to \infty$.

Below T_R the surface height fluctuates within a range of ξ. This means that the steps of a length of order ξ can be thermally generated. The step energy β is then of the order $k_B T/\xi$ as

$$\beta \sim \frac{k_B T}{\xi} \sim \exp \left[-\frac{C}{T_R - T} \right]. \tag{A9.30}$$

It approaches zero as well as its all derivatives by T on approaching the roughening point $T \to T_R$: It has an essential singularity at T_R. For $T > T_R$ the correlation length $\xi = \infty$ and $\beta = 0$.

A9.5 Renormalization group theory on roughening

In the variation method (Appendix A9.4), only the lowest order of perturbation U is considered. The renormalization group method takes the effect of higher order perturbation systematically [113, 114, 145, 146].

By coarse graining the two-dimensional space, one gets the Hamiltonian in a continuum space limit as

$$\mathcal{H} = \int^L dx dy \left[\frac{\gamma}{2} (\nabla z)^2 - V \cos \frac{2\pi z}{a} \right] = \int^L dx dy \left[\frac{\gamma a^2}{2} (\nabla h)^2 - V \cos(2\pi h) \right]. \tag{A9.31}$$

Here the lattice parameter a is notified explicitly, and the height $z(x, y)$ is normalized as $h(x, y) = z(x, y)/a$. L is the linear dimension of the system. Coupling constants in (A9.31) are related to the microscopic parameters in (A9.10) as follows: the surface stiffness is here denoted by $\gamma = 2J/a^2$ and the strength of the one-body potential by $V = U/a^2$.

The height $h(\mathbf{r})$ is expanded in Fourier series with wavevectors q between the lower, $q_c = \pi/L$, and the upper, $\Lambda = \pi/a$, cutoffs.

$$h(\mathbf{r}) = \sum_{q_c < q < \Lambda} h(\mathbf{q}) e^{i\mathbf{q}\mathbf{r}} = \sum_{q_c < q < b^{-1}\Lambda} e^{i\mathbf{q}\mathbf{r}} h(\mathbf{q}) + \sum_{b^{-1}\Lambda < p < \Lambda} e^{i\mathbf{p}\mathbf{r}} h(\mathbf{p})$$

$$\equiv h^<(\mathbf{r}) + h^>(\mathbf{r}). \tag{A9.32}$$

In Eq.(A9.32) the height variable is decomposed in the short wavelength component $h^>(r)$ and the long wavelength component $h^<(r)$. The full Hamiltonian consists of a term which contains only long wavelength components, that with only short wavelength components and that with mixed components as

$$
\mathcal{H}(h(q), h(p)) \equiv \mathcal{H}_0^< + \mathcal{H}_0^> + \mathcal{H}_m = \frac{\gamma a^2}{2} \sum_{q_c < q < b^{-1}\Lambda} q^2 |h(q)|^2 + \frac{\gamma a^2}{2} \sum_{b^{-1}\Lambda < p < \Lambda} p^2 |h(p)|^2
$$
$$
- V \int^L dx dy \cos[2\pi(h^< + h^>)]. \tag{A9.33}
$$

The potential term with the coefficient V is denoted here $\mathcal{H}_m$ for short. In the renormalization method, the short wavelength component $h^>(r)$ is integrated out and its effect on the long wavelength component $h^<(r)$ is systematically studied. The renormalized Hamiltonian $\mathcal{H}^<$ contains only the long wavelength variables $h^<(r)$, and is defined by

$$
\exp\left(-\frac{\mathcal{H}^<}{k_B T}\right) \equiv \mathrm{Tr}^> e^{-\mathcal{H}/k_B T} = \Pi_{b^{-1}\Lambda < p < \Lambda} \int_{-\infty}^{\infty} dh(p) \exp\left[-\frac{\mathcal{H}(h(q), h(p))}{k_B T}\right].
$$
$$
\tag{A9.34}
$$

Here $\mathrm{Tr}^>$ is the trace over the short wavelength component $h(p)$. Actually, the renormalized Hamiltonian $\mathcal{H}^<$ is calculated up to the second order of V. The thermal average $\langle \cdots \rangle$ is defined as that by the short wavelength mode with unperturbed Hamiltonian $\mathcal{H}_0^>$ as

$$
\langle \cdots \rangle = \frac{\mathrm{Tr}^> \cdots \exp(-\mathcal{H}_0^> /k_B T)}{\mathrm{Tr}^> \exp(-\mathcal{H}_0^> /k_B T)}. \tag{A9.35}
$$

It is actually a Gaussian average with $\langle h(p) \rangle = 0$ and the height correlation $\langle |h(p)|^2 \rangle = k_B T / \gamma a^2 p^2$.

Average of the term containing the mixed part $\mathcal{H}_m$ is expanded in the cumulant as

$$
\langle \exp\left(-\frac{\mathcal{H}_m}{k_B T}\right) \rangle = \exp\left[-\langle\left(\frac{\mathcal{H}_m}{k_B T}\right)\rangle + \frac{1}{2}\left\{\langle\left(\frac{\mathcal{H}_m}{k_B T}\right)^2\rangle - \langle\left(\frac{\mathcal{H}_m}{k_B T}\right)\rangle^2\right\} + \cdots\right]. \tag{A9.36}
$$

The first order of V can be calculated as

$$
\begin{aligned}
-\frac{\mathcal{H}^{(1)}}{k_B T} &= \frac{V}{k_B T} \int d^2 r \langle \cos 2\pi(h^< + h^>) \rangle \\
&= \frac{V}{k_B T} \int d^2 r \left[\cos(2\pi h^<)\langle\cos(2\pi h^>)\rangle - \sin(2\pi h^<)\langle\sin(2\pi h^>)\rangle\right] \\
&= \frac{V}{k_B T} \int d^2 r \cos(2\pi h^<) \exp\left[-2\pi^2 \langle (h^>)^2 \rangle\right] \\
&= \frac{\bar{V}}{k_B T} \int d^2 r \cos 2\pi h^<(r). \tag{A9.37}
\end{aligned}
$$

Here the effective potential strength $\bar{V}$ is

$$\bar{V} = V \exp\left[-2\pi^2 g(0)\right] = V b^{-\pi k_B T/\gamma a^2} \tag{A9.38}$$

with the height correlation function of the short wavelength defined by

$$g(r) \equiv \langle h^>(r) h^>(0) \rangle = \sum_{b^{-1}\Lambda < p < \Lambda} \langle |h(p)|^2 \rangle e^{ipr} \approx \frac{k_B T}{4\pi^2 \gamma a^2} \int_{b^{-1}\Lambda}^{\Lambda} \frac{d^2 p}{p^2} e^{ipr}$$

$$\sim \frac{k_B T}{2\pi\gamma a^2} J_0(\Lambda r) \ln b, \tag{A9.39}$$

which decays very fast for $r > \pi/\Lambda$ due to the asymptotic behavior of the Bessel function J_0. The second order term $\mathcal{H}^{(2)}$ is calculated as follows:

$$-\frac{\mathcal{H}^{(2)}}{k_B T} = \frac{1}{2}\left(\frac{V}{k_B T}\right)^2 \int d^2r \int d^2r_1 \left[\langle \cos 2\pi h \cdot \cos 2\pi h_1 \rangle - \langle \cos 2\pi h \rangle \langle \cos 2\pi h_1 \rangle\right]$$

$$\approx \frac{1}{2}\left(\frac{V}{k_B T}\right)^2 \int d^2r \int d^2r_1 (\sin 2\pi h^< \sin 2\pi h_1^<)\langle 2\pi h^> \cdot 2\pi h_1^>\rangle$$

$$= \left(\frac{\pi V}{k_B T}\right)^2 \int dr \int dr_1 g(r - r_1) \left[\cos 2\pi(h^< + h_1^<) - \cos 2\pi(h^< - h_1^<)\right], \tag{A9.40}$$

where the notation $h = h(r)$ and $h_1 = h(r_1)$ is used. Since $g(r - r_1)$ is almost zero for $|r - r_1| > \pi/\Lambda$, the integrand gives contribution only for $r_1 \sim r$. The first term produces the higher harmonics $\cos 2\pi(h^< + h_1^<) \approx \cos 4\pi h^<$, and is irrelevant for further discussion. The second term is expanded as

$$\int d^2r_1 J_0(\Lambda|r - r_1|) \cos 2\pi(h^< - h_1^<) \approx const - 4\pi^3 \Lambda^{-4} A \left(\frac{\pi k_B T}{\gamma a^2}\right)(\nabla h^<)^2, \tag{A9.41}$$

where A is a complicated but a smooth function in the region of interest [146]. Eqs. (A9.40) and (A9.41) shows that the second order perturbation gives the renormalization of the surface tension. The final form of the renormalized Hamiltonian has the same form with the original one as

$$\mathcal{H}^< = \int_0^L dx dy \left[\frac{\bar{\gamma} a^2}{2}(\nabla h^<)^2 - \bar{V} \cos(2\pi h^<)\right]$$

$$= \int_0^{L/b} d\bar{x} d\bar{y} \left[\frac{\bar{\gamma} a^2}{2}(\bar{\nabla} h^<)^2 - \bar{V} b^2 \cos(2\pi h^<)\right], \tag{A9.42}$$

where $x = b\bar{x}$ and $y = b\bar{y}$. The coupling constant is renormalized from the original ones, γa^2 and $U = V(\pi/\Lambda)^2$, to

$$\bar{\gamma} a^2 = \gamma a^2 + \frac{2A(2/X)U^2}{\gamma a^2} \ln b \tag{A9.43}$$

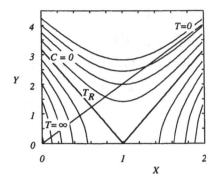

Figure A9.3: The renormalization group flow diagram in $X = 2\gamma a^2/\pi k_B T$ and $Y = 4U/\pi k_B T$ phase space. The temperature variation of a real system takes place along the dashed line. At low temperatures to the right of the separatrix, the potential strength Y increases by renormalization and the surface is smooth. At high temperatures to the left of the separatrix, Y vanishes by renormalization and the surface is rough.

and

$$\bar{U} = \bar{V}\left(\frac{b\pi}{\Lambda}\right)^2 = Ub^{2-\pi k_B T/\gamma a^2} \approx U + U\left(2 - \frac{\pi k_B T}{\gamma a^2}\right)\ln b \qquad (A9.44)$$

with $X = 2\gamma a^2/\pi k_B T$. Therefore, by coarsening the system with the scale $\epsilon = \ln b$, variables X and $Y = 4U/\pi k_B T$ are renormalized as

$$\frac{dX}{d\epsilon} = \frac{A(2/X)}{2}\frac{Y^2}{X} \qquad (A9.45a)$$

$$\frac{dY}{d\epsilon} = 2\left(1 - \frac{1}{X}\right)Y. \qquad (A9.45b)$$

Eqs.(A9.45a,b) are called the renormalization group equations. There is a special fixed point $X^* = 1$, $Y^* = 0$, and it will be shown that the critical behavior is controlled around this point. Therefore, one can assume that the function A is constant with its argument at $X^* = 1$ as $A(2) \equiv A = 0.398$. The flow by renormalization in XY plane is obtained by

$$\frac{dY}{dX} = \frac{4}{A}\frac{X-1}{Y}, \qquad (A9.46)$$

or after integration the renormalization trajectory is obtained by hyperbola as

$$\frac{4}{A}(X-1)^2 - Y^2 = -C. \qquad (A9.47)$$

The flow by renormalization is shown in Fig.A9.3. The line $Y = 0$ for $X \leq 1$ is a line of fixed points. The separatrix $C = 0$ or $\sqrt{A}Y/2 = 1 - X$ corresponds a critical curve, and terminates at the end of fixed points at $X^* = 1$, $Y^* = 0$. Below the separatrix, the strength of the periodic potential U is renormalized to zero, and the surface is rough in a large scale. Since the physical system has fixed values of surface tension $\gamma_0 a^2$ and the potential U_0, the temperature variation occurs along a line $Y/X = 2U_0/\gamma_0 a^2 = const$. This line crosses the critical line $C = 0$ at the roughening temperature

$$k_B T_R = \frac{2}{\pi}\gamma_0 a^2 + \frac{2\sqrt{A}}{\pi}U_0 = \frac{2}{\pi}\gamma_0 a^2 \left[1 + \frac{1}{2}t_c\right] \tag{A9.48}$$

with $t_c = 2\sqrt{A}U_0/\gamma_0 a^2$. Due to the potential U_0 the stiffness γ_0 is modified to the effective one $\gamma_{\text{eff}} = \gamma_0(1 + t_c/2)$, and correspondingly the microscopic coupling $J = \gamma_0 a^2/2$ to the effective one $J_{\text{eff}} = \gamma_{\text{eff}}a^2/2$. The roughening temperature is given as $k_B T_R = 4J_{\text{eff}}/\pi$.

When the temperature is off from T_R, the parameter C is represented as

$$C \approx -\frac{4t(t_c + t)}{A} \tag{A9.49}$$

with $t = (T - T_R)/T_R$. For $T < T_R$, the potential strength Y initially decreases by renormalization until $Y_m = \sqrt{C}$, and then Y increases again. The final increment of Y means the enhancement of the pinning potential U to fix the height to an integer value. The surface is smooth. The correlation length ξ is determined as a length scale until Y reaches to the order unity:

$$\epsilon_c = \ln \xi \approx \int_{Y_0}^1 \frac{X dY}{2(X-1)Y} \approx \int_{\sqrt{C}}^\infty \frac{2dY}{\sqrt{A}Y\sqrt{Y^2 - C}} \approx \frac{\pi}{2\sqrt{|t|t_c}}. \tag{A9.50}$$

Thus the correlation length ξ diverges at the roughening temperature T_R as

$$\xi \approx \exp\left[\frac{\pi}{2}(|t|t_c)^{-1/2}\right]$$
$$\sim \exp\left[\frac{C}{\sqrt{T_R - T}}\right]. \tag{A9.51}$$

The step free energy β is obtained as

$$\beta \sim \xi^{-1} \sim \exp\left[-\frac{C}{\sqrt{T_R - T}}\right]. \tag{A9.52}$$

A9.6 Exact solution of surface roughening: Body-centered solid-on-solid (BCSOS) model

There is an exact solution on the phase transition of the (100) surface of a body-centered cubic (bcc) crystal (Fig.A9.4a). There are eight nearest neighbors for each

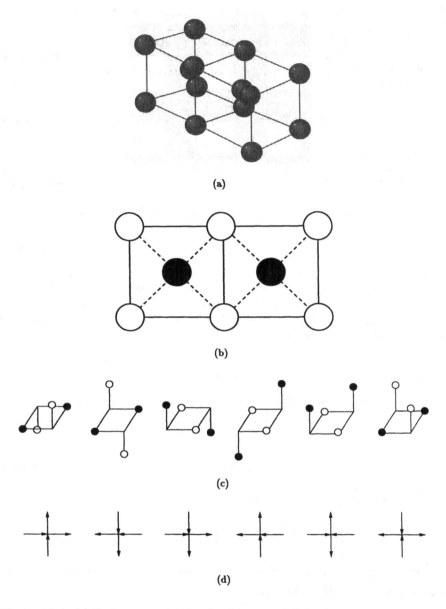

Figure A9.4: (a) Body-centered cubic (bcc) lattice. (b) (100) face of the bcc lattice with the nearest neighbor bonds shown by dashed lines, and the 2nd nearest neighbor by solid lines. (c) Six possible configurations of surface heights. (d) The corresponding six vertices [187].

atom in the bcc crystal. On looking down the (100) face nearest neighbor bonds form a square lattice (Fig.A9.4b). The solid-on-solid model assumes that the surface configuration is described by the height of the column on this square lattice. In a bcc crystal the nearest neighboring heights differ by odd numbers in a unit of half the lattice constant. When the nearest neighbor interaction marked by dashed lines in Fig.A9.4b is attractive and very strong, the height difference is restricted only to unity. Then for a square plaquette there are 6 possible configurations, as shown in Fig.A9.4c. When one draws a vertical line through the middle of an edge, and draw an arrow such that, when one follows the arrow direction, the right stays higher than the left. Then there are six vertex configurations as shown in Fig.A9.4d, and the model is equivalent to the six vertex model [129]. Among four edges from a vertex, two arrows flows inwards and the other two arrows go outwards. The model is originally introduced to explain the residual entropy of ice. In ice, each water molecule is surrounded by four neighboring water molecules by the hydrogen bonds. On the bond connecting two oxygen atoms, a hydrogen atom can be located near one of them. For each oxygen atoms, two hydrogen atoms lie close to it and two others stay away from it. The vertex in Fig.A9.4d represents the oxygen, and the arrow on the edge indicates that the hydrogen lies close to the oxygen. Thus the rule that the number of incoming and the outgoing arrows are the same is called the ice rule [153].

We now impose a second neighbor interaction, $-J_x$ and $-J_y$ in x and y directions respectively, which are shown by solid lines in Fig.A9.4b. The interaction energies $\varepsilon_1 \sim \varepsilon_6$ for vertex configurations 1 to 6 are then given as

$$\varepsilon_1 = \varepsilon_2 = J_y - J_x, \quad \varepsilon_3 = \varepsilon_4 = J_x - J_y, \quad \text{and} \quad \varepsilon_5 = \varepsilon_6 = -J_x - J_y. \quad (A9.53)$$

For an isotropic case, $J_x = J_y = \varepsilon/2 > 0$, the configuration energies are $\varepsilon_1 = \varepsilon_2 = \varepsilon_3 = \varepsilon_4 = 0$ and $\varepsilon_5 = \varepsilon_6 = -\varepsilon$. This is called the F model, and the exact solution is known [129]. The transition temperature is known to be

$$\frac{k_B T_R}{\varepsilon} = (\ln 2)^{-1} \approx 1.4427 \cdots , \quad (A9.54)$$

and the singular part of the surface free energy behaves as

$$F_{\text{sing}} \approx \exp(-\alpha |1 - T/T_R|^{-1/2}), \quad (A9.55)$$

with $\alpha = \pi^2/4\sqrt{\ln 4}$. Step energies in [10] and [11] directions are also known as

$$\frac{E_{[11]}}{\varepsilon} = \frac{2\sqrt{2}}{\sqrt{1-t}} \left\{ \frac{1}{2} \right.$$
$$\left. + \sum_{n=1}^{\infty} (-1)^n [1 + (n-1)\tanh((n-1)\lambda) - n\tanh(n\lambda)] \right\}$$
$$\frac{E_{[10]}}{\varepsilon} = \frac{4}{\sqrt{1-t}} \left\{ \frac{1}{4} \right.$$

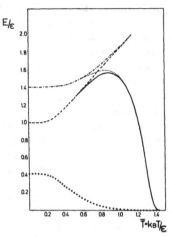

Figure A9.5: Step energies of BCSOS model in two orientations,[10] (solid line) and [11] (dotted line). For comparison, step energies of the Ising model is also shown. Dashed line in [10] direction and dash-dotted line in [11] direction. Crosses represent the difference of the step energies in two directions [187].

$$+ \sum_{n=1}^{\infty} (-1)^n [\frac{1}{2} + (n - \frac{3}{4}) \tanh((n - \frac{3}{4})\lambda) - (n - \frac{1}{4}) \tanh((n - \frac{1}{4})\lambda)] \Big\}$$

$$(A9.56)$$

with

$$\lambda = -\ln t + 2\ln \left(1 + \sqrt{1-t}\right) \qquad (A9.57)$$

and $t = 4\exp(-2\epsilon/k_B T)$. At low temperatures, $k_B T \ll \epsilon$, the step energy is anisotropic, $E_{[11]} > E_{[10]}$. At high temperatures, $k_B T > \epsilon$, it becomes isotropic. Near the roughening temperature where $t \to 1$ and $\lambda \approx 2\sqrt{1-t} \approx 2\sqrt{\ln 4}\sqrt{T_R/T - 1}$, step energies behave as $E_{[11]} \approx E_{[10]} \approx \epsilon\lambda^{-3/2} \exp(-\pi^2/2\lambda) \approx \exp(-\alpha/\sqrt{1 - T/T_R})$. They approach zero as $T \to T_R - 0$, as is shown in Fig.A9.5. The equilibrium shape of facets and the crystalline shape are also obtained [92].

Similar correspondence to the F model is found for the (100) and (110) faces of face centered cubic (fcc) crystal [93].

A14 Advance Rate of a Circular Step

Detailed calculation of the advance rate of a circular step is given in the paper by Burton, Cabrera and Frank [43], and here I reproduce the calculation.

The adatom density or the diffusion field around a circular nucleus with a radius ρ depends only on the radial variable r as $u(r) = c(r) - c_\infty$ due to the symmetry, and the diffusion equation is written as

$$\frac{d^2u}{dr^2} + \frac{1}{r}\frac{du}{dr} - \frac{u}{x_s^2} = 0 \tag{A14.1}$$

in the stationary approximation. The solution which has no singularity at the origin and at infinity is written as

$$u(r) = \begin{cases} u(\rho)\dfrac{I_0(r/x_s)}{I_0(\rho/x_s)} & \text{for } r < \rho \\[3mm] u(\rho)\dfrac{K_0(r/x_s)}{K_0(\rho/x_s)} & \text{for } r > \rho \, . \end{cases} \tag{A14.2}$$

Here I_0 and K_0 are the modified Bessel function of the zeroth order. The ν-th order ones, $I_{\pm\nu}$ and $K_{\pm\nu}$, satisfy the ordinary differential equation

$$z^2\frac{d^2w}{dz^2} + z\frac{dw}{dz} - (z^2 + \nu^2)w = 0, \tag{A14.3}$$

and $I_{\pm\nu}(z)$ is finite at $z \to 0$ and $K_\nu(z)$ is finite at $|z| \to \infty$ and $|\arg(z)| < \frac{\pi}{2}$. For $\nu = 0$, they can be expanded around $z = 0$ as

$$I_0(z) = 1 + \frac{z^2}{4} + \frac{1}{(2!)^2}\left(\frac{z^2}{4}\right)^2 + \cdots \tag{A14.4}$$

$$K_0(z) = -\left[\ln(\frac{z}{2}) + \gamma\right]I_0(z) + \frac{1}{(1!)^2}\frac{z^2}{4} + (1 + \frac{1}{2})\frac{1}{(2!)^2}\left(\frac{z^2}{4}\right)^2 + \cdots \tag{A14.5}$$

with Euler's constant $\gamma = 0.57721\,56649\cdots$. Their asymptotic forms for $z \to \infty$ are

$$I_0(z) \sim \frac{e^z}{\sqrt{2\pi z}}\left[1 + \frac{1}{8z} + \cdots\right] \tag{A14.6}$$

$$K_0(z) \sim \sqrt{\frac{\pi}{2z}}e^{-z}\left[1 - \frac{1}{8z} + \cdots\right]. \tag{A14.7}$$

By differentiation, the following relations hold;

$$I_0'(z) = I_1(z), \quad K_0'(z) = -K_1(z), \quad I_0(z)K_1(z) + I_1(z)K_0(z) = \frac{1}{z}. \tag{A14.8}$$

The local equilibrium is expected around a circular nucleus;

$$u(\rho) = c_{eq}(\rho) - c_\infty. \tag{A14.9}$$

On the other hand, the advance velocity of the step is derived from the material conservation as

$$
\begin{aligned}
v(\rho) &= D_s a^2 \left. \frac{\partial c}{\partial r} \right|_{\rho-}^{\rho+} = D_s a^2 \frac{u(\rho)}{x_s} \left[\frac{K_0'(\rho/x_s)}{K_0(\rho/x_s)} - \frac{I_0'(\rho/x_s)}{I_0(\rho/x_s)} \right] = -D_s a^2 \frac{u(\rho)}{x_s} \frac{K_1 I_0 + I_1 K_0}{K_0 I_0} \\
&= -D_s a^2 \frac{u(\rho)}{x_s} \frac{1}{\rho/x_s} \frac{1}{K_0 I_0} = -\frac{D_s a^2}{\rho} \frac{1}{K_0(\rho/x_s) I_0(\rho/x_s)} u(\rho).
\end{aligned} \tag{A14.10}
$$

When the radius ρ is larger than x_s, the approximation $K_0(\rho/x_s) I_0(\rho/x_s) \approx x_s/2\rho$ holds from Eq.(A14.6) and (A14.7), and $v(\rho)$ is obtained as

$$v(\rho) = \frac{D_s a^2}{\rho} \frac{1}{x_s/2\rho} (c_\infty - c_{eq}(\rho)) = \frac{2 D_s a^2}{x_s} \left(c_\infty - c_{eq} - c_{eq} \frac{\beta a^2}{\rho k_B T} \right). \tag{A14.11}$$

For the straight step with $\rho \to \infty$, the advance velocity reduces to v_0 as

$$v(\rho \to \infty) = \frac{2 D_s a^2}{x_s} (c_\infty - c_{eq}) = v_0. \tag{A14.12}$$

Then the velocity of a circular step is approximated as

$$v(\rho) = v_0 \left(1 - \frac{\rho_c}{\rho} \right), \tag{A14.13}$$

where ρ_c is the critical radius of the two-dimensional nucleation defined in (6.23) as

$$\rho_c = \frac{c_{eq}}{c_\infty - c_{eq}} \frac{\beta a^2}{k_B T} = \frac{\beta a^2}{\Delta \mu}. \tag{A14.14}$$

A15 Advance Velocity of a Spiral Step

There are various ways to represent the spiral curve. One way is to use an arc length s along the spiral and the angle θ of the normal vector $\mathbf{n}$ from the y axis as shown in Fig.A15.1a.

$$\theta = \theta(s). \tag{A15.1}$$

From Fig.A15.1a it is obvious that the radius of the curvature ρ is related to the variation of the arc length ds and the associated angle change $d\theta$ as $ds = \rho d\theta$, and the curvature is given as

$$\kappa = \frac{1}{\rho} = \frac{d\theta}{ds}. \tag{A15.2}$$

The same spiral can be represented by the polar coordinate (r, ϕ) with its origin at the spiral center as

$$\phi = \phi(r). \tag{A15.3}$$

Figure A15.1b shows that the variation of an arc length ds is written as

$$ds = \sqrt{(dr)^2 + (rd\phi)^2} = dr\sqrt{1 + (r\phi')^2}, \tag{A15.4}$$

where the prime means the derivative by r: $' = d/dr$. There is still another possibility of representing the spiral by the angle ψ between the radial vector $\mathbf{r}$ and the tangential vector $\mathbf{t}$ of the spiral (Fig.A15.1c) [44]:

$$\psi = \psi(r). \tag{A15.5}$$

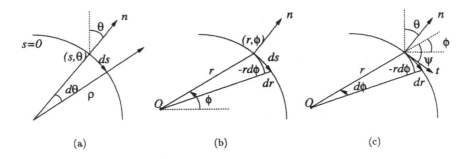

(a)   (b)   (c)

Figure A15.1: Three ways of representing a curved spiral. (a) By arc length s and the angle θ of the normal vector $\mathbf{n}$ making with y axis. (b) The polar coordinate (r, ϕ). (c) The length r of the radial vector $\mathbf{r}$ and the angle ψ of the tangential vector $\mathbf{t}$ and $\mathbf{r}$. Angles are related as $\psi = \theta + \phi$.

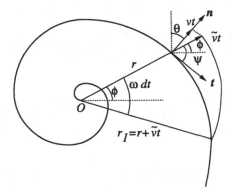

Figure A15.2: A spiral steadily rotating with an angular velocity ω. $\tilde{v}$ is the local velocity in the radial direction, and v is velocity normal to the spiral.

As is clear from Fig.A15.1c, the three angles are related as

$$\psi = \phi + \theta. \tag{A15.6}$$

Also from Fig.A15.1c it is evident that the relation

$$r\phi' = -\tan\psi \tag{A15.7}$$

holds. Then the curvature κ is written as

$$\kappa = \frac{1}{\rho} = \frac{d\theta}{ds} = \frac{d\psi - d\phi}{dr\sqrt{1 + (r\phi')^2}} = \frac{\sin\psi}{r} + \psi'\cos\psi. \tag{A15.8}$$

When the spiral is advancing steadily, the spiral looks as if it is rotating steadily. By denoting the radial velocity at a point (r, ϕ) as $\tilde{v}$, the point on the curve at a polar angle ϕ moves to

$$r_1 = r(t) + \tilde{v}dt \tag{A15.9}$$

after a time dt. Since the movement looks like a steady rotation with an angular velocity ω, the angle $\phi(r_1, t)$ on a spiral at time t should rotate to $\phi(r, t)$ after a time elapse dt:

$$\phi(r_1, t) + \omega dt = \phi(r, t). \tag{A15.10}$$

(See Fig.A15.2). Inserting (A15.9) into (A15.10) and expanding in terms of a small variable dt, radial velocity $\tilde{v}$ is written as

$$\tilde{v} = -\frac{\omega}{\phi'}. \tag{A15.11}$$

The normal component of the step velocity should be

$$v(r) = \tilde{v} \cos\left(\phi + \theta - \frac{\pi}{2}\right) = -\frac{\omega}{\phi'} \sin\psi, \qquad (A15.12)$$

and this velocity should coincide with the velocity given by the step curvature as

$$v(r) = v_0 \left(1 - \frac{\rho_c}{\rho}\right). \qquad (A15.13)$$

From Eq.(A15.7-A15.8) and (A15.12-A15.13), the spiral $\psi = \psi(r)$ should satisfy the first order differential equation

$$\frac{d\psi}{dr} = -\frac{\tan\psi}{r} + \frac{1}{\rho_c \cos\psi} - \frac{\omega}{v_0 \rho_c} r. \qquad (A15.14)$$

Boundary conditions are set at $r \to 0$ and at $r \to \infty$: At $r \to \infty$, the spiral looks like a circle and the angle $\psi \to \pi/2$. At $r \to 0$, the radius of the curvature decreases, but it should be larger than ρ_c, because if the radius is smaller than ρ_c, the crystal melt back as is evident from Eq.(A15.13). Therefore, at $r = 0$ the curvature is ρ_c. From (A15.8) ψ should approach zero as $r \to 0$ in order to obtain a finite κ. Then $\sin\psi/r \sim \psi/r \sim d\psi/dr$, and at $r \to 0$ the relation $1/\rho_c \sim 2d\psi/dr$ holds. Therefore, $\psi \sim r/2\rho_c$ for small r.

Since the angle ψ satisfies the first order differential equation (A15.16) the solution contains only one integral constant. Thus two boundary conditions cannot be satisfied in general. The solution exists only for some special values of ω. Thus, Eq.(A15.14) is the eigenvalue equation to determine the eigenvalue ω. From the numerical calculation, the eigenvalue is obtained to be

$$\omega \approx 0.33 v_0/\rho_c. \qquad (A15.15)$$

The period T for one turn of the spiral is $T = 2\pi/\omega$. Since the advance velocity of the step is v_0 asymptotically at $r \to \infty$, the step separation $\lambda = v_0 T$ is calculated as

$$\lambda = \frac{2\pi v_0}{\omega} \approx 19\rho_c. \qquad (A15.16)$$

The step separation of a steadily growing spiral is 19 times larger than the critical radius ρ_c of the two-dimensional nucleation.

From the relation (A15.7), the polar angle ϕ for $r \to 0$ is determined from the relation $r\phi' \approx -\psi \approx -r/2\rho_c$ to be

$$\phi \approx -\frac{r}{2\rho_c}. \qquad (A15.17)$$

This is the Archimedes's spiral turning right. The spiral turning left is written as $\phi = r/2\rho_c$, This is used to obtain the formula (15.1).

A21 Parabolic Coordinate

The Cartesian coordinate (x, y, z) and the parabolic coordinate (ξ, η, ϕ) are related as

$$x = \sqrt{\xi\eta}\cos\phi, \quad y = \sqrt{\xi\eta}\sin\phi, \quad z = \frac{1}{2}(\eta - \xi), \tag{A21.1}$$

where $r = \sqrt{x^2 + y^2 + z^2} = \frac{1}{2}(\eta + \xi)$. Inverse relation is written as

$$\xi = r + z, \quad \eta = r - z, \quad \phi = \arctan\frac{y}{x}. \tag{A21.2}$$

The line element is represented in both coordinate as

$$(ds)^2 = (dx)^2 + (dy)^2 + (dz)^2 = (h_\eta d\eta)^2 + (h_\xi d\xi)^2 + (h_\phi d\phi)^2 \tag{A21.3}$$

with

$$h_\eta = \sqrt{\frac{\eta + \xi}{4\eta}}, \quad h_\xi = \sqrt{\frac{\eta + \xi}{4\xi}}, \quad h_\eta = \sqrt{\eta\xi}. \tag{A21.4}$$

The gradient vector of the field is

$$\begin{aligned} \nabla u &= \frac{\partial u}{\partial x}\hat{x} + \frac{\partial u}{\partial y}\hat{y} + \frac{\partial u}{\partial z}\hat{z} \\ &= \sqrt{\frac{4\eta}{\eta + \xi}}\frac{\partial u}{\partial \eta}\hat{\eta} + \sqrt{\frac{4\xi}{\eta + \xi}}\frac{\partial u}{\partial \xi}\hat{\xi} + \sqrt{\frac{1}{\eta\xi}}\frac{\partial u}{\partial \phi}\hat{\phi} \end{aligned} \tag{A21.5}$$

with basis vectors $(\hat{x}, \hat{y}, \hat{z})$ and $(\hat{\xi}, \hat{\eta}, \hat{\phi})$ in each coordinate system. The basis vector $\hat{z}$ is related to those of parabolic coordinate as

$$\hat{z} = \frac{1}{2h_\eta}\hat{\eta} - \frac{1}{2h_\xi}\hat{\xi}. \tag{A21.6}$$

The Laplacian is written as

$$\Delta u = \frac{4}{\eta + \xi}\left[\frac{\partial}{\partial \eta}\left(\eta\frac{\partial u}{\partial \eta}\right) + \frac{\partial}{\partial \xi}\left(\xi\frac{\partial u}{\partial \xi}\right)\right] + \frac{1}{\eta\xi}\frac{\partial^2 u}{\partial \phi^2}. \tag{A21.7}$$

By denoting the interface as $\eta = \eta_i(\xi, \phi)$, the unnormalized tangential vectors are written as

$$\mathbf{t}_1 = h_\eta\frac{\partial \eta_i}{\partial \xi}\hat{\eta} + h_\xi\hat{\xi} \tag{A21.8}$$

$$\mathbf{t}_2 = h_\eta\frac{\partial \eta_i}{\partial \phi}\hat{\eta} + h_\phi\hat{\phi}. \tag{A21.9}$$

The normal vector is then obtained as

$$\mathbf{n} = N\left[\frac{1}{h_\eta}\hat{\eta} - \frac{1}{h_\xi}\frac{\partial \eta_i}{\partial \xi}\hat{\xi} - \frac{1}{h_\phi}\frac{\partial \eta_i}{\partial \phi}\hat{\phi}\right] = N\left[\sqrt{\frac{4\eta}{\eta + \xi}}\hat{\eta} - \sqrt{\frac{4\xi}{\eta + \xi}}\frac{\partial \eta}{\partial \xi}\hat{\xi} - \sqrt{\frac{1}{\eta\xi}}\frac{\partial \eta}{\partial \phi}\hat{\phi}\right] \tag{A21.10}$$

with N being determined from the normalization condition $|\mathbf{n}| = 1$.

The interface velocity of the dendrite in laboratory frame is written as $\mathbf{v} = V\hat{z} + \dot{\eta}h_\eta\hat{\eta}$. By using $\hat{z}$ and $\mathbf{n}$ given in Eq.(A21.6) and (A21.10), one obtains the interface normal velocity as

$$v_n = N\left[V\left(\frac{1}{2h_\eta^2} + \frac{1}{2h_\xi^2}\frac{\partial\eta_i}{\partial\xi}\right) + \dot{\eta}\right], \qquad (A21.11)$$

which is explicitly presented in Eq.(21.3).

A24 Dendritic Growth

A24.1 Boundary integral equation for the diffusion-controlled growth

In a steady state approximation, the diffusion equation in the moving frame is written as

$$\hat{L}u \equiv \nabla^2 u + \frac{2}{l_D}\frac{\partial u}{\partial z'} = 0, \tag{A24.1}$$

where z' is a new coordinate $z' = z - vt$ and $l_D = 2D/v$ is the diffusion length. Hereafter in this section we drop the prime on z. By using the adjoint operator $\hat{L}^\dagger$ defined by

$$\hat{L}_1^\dagger g(\mathbf{r}, \mathbf{r}_1) \equiv \nabla_1^2 g - \frac{2}{l_D}\frac{\partial g}{\partial z_1}, \tag{A24.2}$$

the Green's theorem says that the integral over the region Ω_1 enclosed by the periphery Γ_1 satisfies the relation:

$$\int d\Omega_1 \left[g(\mathbf{r}, \mathbf{r}_1)\hat{L}_1 u(\mathbf{r}_1) - u(\mathbf{r}_1)\hat{L}_1^\dagger g(\mathbf{r}, \mathbf{r}_1) \right]$$
$$= \oint d\Gamma_1 \left[g(\mathbf{r}, \mathbf{r}_1)\frac{\partial u(\mathbf{r}_1)}{\partial n_1} - u(\mathbf{r}_1)\frac{\partial g(\mathbf{r}, \mathbf{r}_1)}{\partial n_1} + \frac{2}{l_D}n_{1,z}g(\mathbf{r}, \mathbf{r}_1)u(\mathbf{r}_1) \right]. \tag{A24.3}$$

When $u(\mathbf{r}_1)$ satisfies the diffusion equation (A24.1) and g is a Green's function of the operator $\hat{L}_1^\dagger$ as

$$\hat{L}_1^\dagger g(\mathbf{r}, \mathbf{r}_1) = -\delta(\mathbf{r} - \mathbf{r}_1), \tag{A24.4}$$

then (A24.3) reduces to the integral equation over the crystal-liquid interface Γ_{SL}.

$$\int d\Gamma_{1,SL}g(\mathbf{r}, \mathbf{r}_1)\frac{\partial u}{\partial n_1} = \int d\Gamma_{1,SL}h(\mathbf{r}, \mathbf{r}_1)u(\mathbf{r}_1). \tag{A24.5}$$

Here we have used the boundary condition that far from the liquid-crystal interface, the diffusion field and its derivative vanish: $u(r \to \infty) = \partial u/\partial n|_{r \to \infty} = 0$. The Green's function g is obtained by the Fourier transformation as

$$g(\mathbf{r}, \mathbf{r}_1) = \int \frac{d^2\mathbf{q}}{(2\pi)^2}\frac{e^{-i\mathbf{q}(\mathbf{r} - \mathbf{r}_1)}}{q_x^2 + (q_z - il_D^{-1})^2 + l_D^{-2}} = \frac{1}{2\pi}e^{-(\zeta - \zeta_1)/l_D}K_0\left(\frac{|\mathbf{r} - \mathbf{r}_1|}{l_D}\right) \tag{A24.6}$$

with K_0 being the modified Bessel function explained in Appendix A14. Another integral kernel h is obtained as

$$h(\mathbf{r}, \mathbf{r}_1) = \frac{\partial g}{\partial n_1} - \frac{2}{l_D}n_{1,z}g - c(\mathbf{r}_1)\delta(\mathbf{r} - \mathbf{r}_1). \tag{A24.7}$$

The coefficient c enters because the position $\mathbf{r}$ in Eq.(A24.5) lies on the boundary Γ_{SL}. Instead of determining $c(\mathbf{r})$ directly, one can impose a sum rule on h as

$$\int d\Gamma_{1,SL} h(\mathbf{r},\mathbf{r}_1) = -1. \tag{A24.8}$$

The condition is easily derived by inserting the trivial solution $u = const$ of the diffusion equation in Eq.(A24.3) [30, 165]. Equation (A24.5) relates the diffusion field u to its normal derivative $\partial u/\partial n$ at the liquid-crystal interface. This relation is used in the numerical simulation described in Appendix A24.3.

If there is a diffusion in the crystal with the diffusion constant D_S, we can obtain another boundary integral equation similar to Eq.(A24.5) with the Green's function g_S and an integral kernel h_S. In them, the diffusion length is replaced by the diffusion length in the solid $l_{D,S} = 2D_S/v$, and in h_S the normal vector is replaced by $\mathbf{n}_S = -\mathbf{n}$. Due to the difference of the direction of the normal vector, the sum rule for h_S is different from (A24.8) and is written as [168]

$$\int d\Gamma_{1,SL} h_S(\mathbf{r},\mathbf{r}_1) = 0. \tag{A24.9}$$

The diffusion field is continuous at the boundary; $u_S = u$. When the heat is conducted in the crystal as well as in the liquid, the energy conservation boundary condition (18.7) is altered as

$$v_n = -\left(D\frac{\partial u}{\partial n} - D_S\frac{\partial u_S}{\partial n}\right). \tag{A24.10}$$

In a symmetric model where $D_S = D$, one gets simple relations; $l_{D,S} = l_D$, $g_S = g$ and $h - h_S = -\delta(\mathbf{r}-\mathbf{r}_1)$. By subtracting Eq.(A24.5) from the corresponding one in the crystal, and by utilizing the above relations and the energy conservation Eq.(A24.10), one gets

$$\int d\Gamma_{1,SL} g(\mathbf{r},\mathbf{r}_1)\left[\frac{\partial u_S}{\partial n_1} - \frac{\partial u}{\partial n_1}\right] = \frac{1}{D}\int d\Gamma_{1,SL} g v_n$$
$$= \int d\Gamma_{1,SL}\left[h_S(\mathbf{r},\mathbf{r}_1) - h(\mathbf{r},\mathbf{r}_1)\right] u(\mathbf{r}_1) = u(\mathbf{r}). \tag{A24.11}$$

When the crystal grows steadily in z direction with the velocity $v\hat{z}$, the normal velocity is $v_n = vn_z$ and the integration can be transformed as $\int d\Gamma_{1,SL} n_z = \int dx_1$. Also by imposing the local equilibrium condition (18.11) or

$$u(\mathbf{r}) = \Delta - d\kappa, \tag{A24.12}$$

one obtains a simple equation

$$\Delta - d\kappa = \frac{2}{l_D}\int_{-\infty}^{\infty} dx_1 g(\mathbf{r},\mathbf{r}_1) \tag{A24.13}$$

for the symmetric model. This is the relation often used in the theoretical analysis, as will be explained in Appendix A24.2.

A24.2 Microscopic Solvability Theory of Dendritic Growth

The solvability theory is very mathematical and its detailed explanation is out of scope of the present notes. I sketch here briefly its main idea and the result [124, 105, 33, 156].

When the crystal is growing steadily, $\partial\zeta/\partial t = 0$, and the diffusion constants in solid and melt is the same (symmetric model), the profile $z = \zeta(x)$ of a two-dimensional crystal is determined by (A24.13) or

$$\Delta - d\kappa = \frac{1}{\pi l_D} \int_{-\infty}^{\infty} dx_1 e^{-(\zeta(x)-\zeta(x_1))/l_D} K_0 \left(\frac{\sqrt{(x-x_1)^2 + (\zeta(x)-\zeta(x_1))^2}}{l_D} \right). \quad (A24.14)$$

Here K_0 is the modified Bessel function of the zeroth order, explained in Appendix A14. Without capillarity ($d = 0$), Ivantsov parabola $\zeta_{IV}(x) = -x^2/2\rho$ satisfies Eq.(A24.14) at the supercooling Δ [154]. With the capillarity, the interface $\zeta(x)$ deviates slightly from the Ivantsov parabola, ζ_{IV}. By using the Ivantsov relation between the supercooling Δ and the profile ζ_{IV} and measuring the length in unit of the tip radius ρ of the Ivantsov parabola, one gets the relation

$$\sigma A(x,\zeta)\kappa = \Gamma_2(P, x, \zeta) - \Gamma_2(P, x, \zeta_{IV}). \quad (A24.15)$$

Here $P = \rho/l_D$ is the Peclet number, the function Γ_2 is defined by

$$\Gamma_2(P, x, \zeta) = \frac{1}{\pi} \int_{-\infty}^{\infty} dx_1 e^{-P(\zeta(x)-\zeta(x_1))} K_0 \left(P\sqrt{(x-x_1)^2 + (\zeta(x)-\zeta(x_1))^2} \right), \quad (A24.16)$$

and σ is the stability parameter defined by

$$\sigma = \frac{d_0}{\rho P} = \frac{2D d_0}{\rho^2 v} = \frac{d_0 v}{2D P^2}. \quad (A24.17)$$

The surface stiffness is assumed to have four-fold symmetry with the anisotropy factor A defined as

$$A(x,\zeta) = 1 - \epsilon \cos 4\theta = 1 - \epsilon + \epsilon \frac{8(\partial\zeta/\partial x)^2}{[1+(\partial\zeta/\partial x)^2]^2}. \quad (A24.18)$$

ϵ here means the strength of the anisotropy. Since σ is proportional to the small capillary length d_0, it is also small. But the usual perturbation expansion in σ is not applicable, since σ couples with the highest derivative of the height or the curvature

$$\kappa = -\frac{\partial^2\zeta(x)/\partial x^2}{[1+(\partial\zeta/\partial x)^2]^{3/2}}. \quad (A24.19)$$

One has to use singular perturbation method [16]. The situation is similar to the quantum mechanics [171]: In the Schrödinger equation

$$-\frac{\hbar^2}{2m}\frac{\partial^2\Psi}{\partial x^2} - V(x)\Psi = E\Psi, \quad (A24.20)$$

the Planck constant $h = 2\pi\hbar$ acts as a small parameter, but it is coupled to the highest second derivative of the wave function Ψ. The approximation to the wave function Ψ can be obtained by a singular perturbation, known as WKB approximation [171].

Up to the linear order of the deviation of the interface profile from the Ivantsov parabola

$$\zeta(x) - \zeta_{\mathrm{IV}}(x) \equiv (1 + x^2)^{3/4} Z(x), \tag{A24.21}$$

Eq.(A24.15) is transformed to the following linear and inhomogeneous integro-differential equation [124]

$$
\begin{aligned}
\mathcal{L}Z &\equiv \sigma \frac{d^2 Z(x)}{dx^2} + \frac{(1 + x^2)^{1/2}}{A(x)} Z(x) \\
&\quad + \frac{(1 + x^2)^{3/4}}{2\pi A} \mathcal{P} \int dx_1 \frac{(x + x_1)(1 + x_1^2)^{3/4}}{(x - x_1)[1 + \frac{1}{4}(x + x_1)^2]} Z(x_1) \\
&= \frac{\sigma}{(1 + x^2)^{3/4}}.
\end{aligned}
\tag{A24.22}
$$

Here $\mathcal{P}$ denotes the principal value. For the existence of a nontrivial solution, the inhomogeneous term in the r.h.s. should be orthogonal to the zero eigenvector $\tilde{Z}(x)$ of the adjoint operator $\mathcal{L}^\dagger$ as

$$\Theta(\sigma, \epsilon) \equiv \int_{-\infty}^{\infty} dx \frac{\tilde{Z}(x; \sigma, \epsilon)}{(1 + x^2)^{3/4}} = 0. \tag{A24.23}$$

This equation is an eigenvalue equation to determine the eigenvalue σ. $\tilde{Z}(x)$ is obtained by the WKB method, and $\Theta(\sigma, \epsilon)$ is calculated to behave approximately as

$$\Theta(\sigma, \epsilon) \sim \exp\left(-\frac{C}{\sqrt{\sigma}}\right) \cos\left(\frac{\pi}{2} \sqrt{\frac{\sigma_0 \epsilon^{7/4}}{\sigma}}\right) \tag{A24.24}$$

at small σ and ϵ. This equation shows that with an isotropic surface tension ($\epsilon = 0$), Θ can never be 0 for finite σ or finite velocity. To realize steady growth, an anisotropy ϵ is necessary in the surface stiffness. For $\epsilon \neq 0$ there are infinitely many but discrete numbers of stationary solutions

$$\sigma \approx \sigma_0 \epsilon^{7/4} (1 + 2n)^2 \quad (\text{with } n = 0, 1, 2, \cdots, \infty). \tag{A24.25}$$

From the linear stability analysis around these stationary solutions, the steady state with the minimum σ or the maximum velocity v is found to be stable, but the other solutions are unstable.

The dendrite tip radius of the curvature ρ and the growth velocity v thus satisfy the scaling relation

$$v\rho^2 = \frac{2Dd_0}{\sigma} = \frac{2Dd_0}{\sigma_0} \epsilon^{-7/4} \tag{A24.26}$$

as well as the Ivantsov relation (21.13), which reflects the total energy conservation and is influenced little by capillarity. The tip radius and the velocity is then determined uniquely for a given supercooling Δ as

$$\rho = \frac{d_0}{\sigma_0} \frac{\epsilon^{-7/4}}{P(\Delta)} \sim d_0 \frac{\pi}{\sigma_0} \Delta^{-2} \epsilon^{-7/4} \tag{A24.27}$$

and

$$v = \frac{2D\sigma_0}{d_0} P^2(\Delta) \epsilon^{7/4} \sim \frac{2D\sigma_0}{\pi^2 d_0} \Delta^4 \epsilon^{7/4}. \tag{A24.28}$$

In the last expression in Eq.(A24.27) and (A24.28) Peclet number $P(\Delta)$ is approximated by its limit for small supercooling in two dimensions.

A24.3 Numerical Simulation of the Dendritic Profile

Since the dendritic pattern is quite different from the flat interface, linear and weakly nonlinear analysis is insufficient. To find out the strongly nonlinear and far from nonequilibrium pattern, numerical simulation is a very useful method. There are various different methods to simulated the profile of the crystal [135, 102, 103, 165, 109, 110, 87]. The main problem of the simulation is the sharpness of the interface. If one solves the bulk diffusion equation numerically, a discrete mesh is necessary. When the interface moves, it will be off from the mesh points. Since the interface is susceptible to the fluctuation because of the Mullins-Sekerka instability, one has to be careful in tracing the interface motion. Also, from the microscopic solvability analysis, the anisotropy in the system is said to be essential to stabilize the dendrite tip. By using the mesh for the diffusion field, an artificial anisotropy from the mesh lattice can be introduced. There are various simulation methods to circumvent these obstacles. First we briefly summarize ideas of some proposed methods.

By assuming the steady growth of the dendrite, the diffusion equation can be transformed to the integro-differential equation (A24.5) with the help of the Green's theorem, as explained in Appendix A24.1. By solving this interface equation numerically there occur no problems stated above, since grid points lie on the interface. By integrating along one side of a steadily growing dendrite from the tail to the tip, one can search the selected velocity such as to produce a round tip which connects smoothly to the other side of the dendrite symmetrically [135, 102, 103]. This is the numerical realization of the solvability criterion of a symmetric dendrite in a stationary code. Our simulation [165] relies on the integro-differential equation (A24.5), but it allows the variation of the system velocity so as to select the final profile spontaneously. There are other methods which don't use boundary integral equation. In a phase field model, the interface is defined as the domain boundary of the phase field and is treated as a diffuse object, in order to get rid of the difficulty associated to the sharp interface [109, 110, 111, 117]. This is so far the sole model on which a three-dimensional simulation is performed [109, 111]. The phase-field model, however, uses a mesh to solve the diffusion equation, and the mesh-lattice anisotropy is expected to affect the quantitative results. Recently a new algorithm with multiple meshes are proposed to solve the diffusion equation. These meshes are mutually shifted translationally and rotationally [87]. With this multiple-mesh method the real dynamical evolution of the crystal interface is simulated in many systems.

Here I describe our boundary integral algorithm applied first to study the selection problem of a two-dimensional dendrite profile of a one-sided model, where the diffusion takes place only in the liquid phase [165]. The time evolution of the crystal-liquid interface is traced by solving the quasi-stationary distribution of the diffusion field by the boundary element method [30]. First assume that the profile $\Gamma_{\mathrm{SL}}(t)$ and the frame velocity $v(t)$ at time t are given. Then modification of Γ_{SL} and v to the asymptotic form is performed in the following processes.

1. Calculate the diffusion field u at the boundary from the local equilibrium condition (A24.12),

2. calculate the Green's function g and the integral kernel h by using the frame velocity v,

3. solve the linear equation (A24.5) for the normal gradient $q = \partial u/\partial n$, since g, h, u and Γ_{SL} are known,

4. the normal velocity is calculated by the conservation condition as $\overset{\bullet}{v}_n(\mathbf{r}) = -Dq(\mathbf{r})$,

5. then evolve the crystal profile in the normal direction $\mathbf{n}$ by $\mathbf{r}(t+dt) = \mathbf{r}(t)+v_n\mathbf{n}dt$ to a new configuration $\Gamma_{\mathrm{SL}}(t + dt)$ at time $t + dt$,

6. adjust the moving frame velocity v to the tip velocity of the dendrite v_{tip} by a rather ad hoc relaxation $v(t + dt) = v(t) - (v(t) - v_{\mathrm{tip}}(t))dt/\tau$,

7. and then iterate back to 1. and repeat the whole procedure until the steady state is realized.

If the steady state is realized, the ad hoc relaxation of the frame velocity by the procedure 6. does not affect the resulting asymptotics. To solve the linear equation (A24.5) numerically by computer, one has to discretize the crystal front into a polygon, whose corner points are denoted by $\mathbf{r}_j$ with $j = 1, \cdots, N$. The diffusion field u and the normal gradient $q = \partial u/\partial n$ at a corner point $\mathbf{r}_j$ are denoted by u_j and q_j respectively. Every quantity at a point $\mathbf{r}$ on a polygon edge Γ_j between $\mathbf{r}_j$ and $\mathbf{r}_{j+1}$ is interpolated linearly, for example, for u as

$$u(\mathbf{r}) = \phi_1(\xi)u_j + \phi_2(\xi)u_{j+1} \tag{A24.29}$$

and for the position $\mathbf{r}$ as

$$\mathbf{r} = \phi_1(\xi)\mathbf{r}_j + \phi_2(\xi)\mathbf{r}_{j+1} \tag{A24.30}$$

with an interpolation functions [30]

$$\phi_1(\xi) = \frac{1-\xi}{2} \qquad \phi_2(\xi) = \frac{1+\xi}{2} \tag{A24.31}$$

with $-1 \le \xi \le 1$. The integration along the edge Γ_j is written as

$$\int_{\mathbf{r}_j}^{\mathbf{r}_{j+1}} d\Gamma_j = \frac{s_j}{2} \int_{-1}^{1} d\xi \tag{A24.32}$$

with $s_j = |\mathbf{r}_{j+1} - \mathbf{r}_j|$ being the length of the edge segment Γ_j. Then the integral equation (A24.5) can be written as a matrix equation

$$\sum_{j=1}^{N} G_{ij} q_j = \sum_{j=1}^{N} H_{ij} u_j, \tag{A24.33}$$

where

$$G_{ij} = \frac{s_j}{2} \int_{-1}^{1} d\xi \, g \, [\mathbf{r}_j + \phi_2(\xi)\mathbf{s}_j, \mathbf{r}_i] \, \phi_1(\xi) + \frac{s_{j-1}}{2} \int_{-1}^{1} d\xi \, g \, [\mathbf{r}_j - \phi_1(\xi)\mathbf{s}_{j-1}, \mathbf{r}_i] \, \phi_2(\xi) \tag{A24.34}$$

and

$$H_{ij} = \frac{s_j}{2} \int_{-1}^{1} d\xi \, h \, [\mathbf{r}_j + \phi_2(\xi)\mathbf{s}_j, \mathbf{r}_i] \, \phi_1(\xi) + \frac{s_{j-1}}{2} \int_{-1}^{1} d\xi \, h \, [\mathbf{r}_j - \phi_1(\xi)\mathbf{s}_{j-1}, \mathbf{r}_i] \, \phi_2(\xi) \tag{A24.35}$$

with g and h given in Eqs.(A24.6) and (A24.7), respectively. The sum rule (A24.8) reduces to the condition

$$\sum_{j=1}^{N} H_{ij} = -1 \tag{A24.36}$$

and the diagonal element H_{ii} is determined from

$$H_{ii} = -\sum_{j \ne i} H_{ij} - 1. \tag{A24.37}$$

The integration in Eq.(A24.34-A24.35) are performed by the four-point Gaussian quadrature method [1, 30]. When the point $\mathbf{r}_i$ happens to be one of the end points of the linear segment Γ_j where the integration is performed, the Green's function g contains the logarithmic singularity as

$$g(r, r_1) \approx \exp\left(-\frac{\zeta - \zeta_1}{l_D}\right) \left[-\ln\left(\frac{|\mathbf{r} - \mathbf{r}_1|}{2l_D}\right) - \gamma\right] \tag{A24.38}$$

with Euler constant γ. Integration then uses the Gaussian integration with a logarithmic singularity [1, 30].

Since our main interest lies in the evolution of the dendrite tip, we divide the interface into three parts along z: a tip region, a transition region and a tail region.

Evolution of the tip region is treated fully as described. The tail region consists of an Ivantsov parabola appropriate to the steady growth assumption with the frame velocity v. A transition region connects the tip and tail regions smoothly.

Extension of the method to include the kinetic effect or dendrite tilting is straight-forward. When the kinetic coefficient $K(\mathbf{n})$ is finite and anisotropic, the local equilibrium assumption does not hold any more. Here one has to use the kinetic condition (18.8) [50]. In the steady state the diffusion field at the interface $u(\mathbf{r}_1)$ is decomposed as

$$u(\mathbf{r}_1) = \Delta - d\kappa - \frac{v_n}{K(\mathbf{n})} = u_0(\mathbf{r}_1) + \frac{D}{K(\mathbf{n})}\frac{\partial u}{\partial n} \qquad (A24.39)$$

with

$$u_0(\mathbf{r}_1) = \Delta - d\kappa. \qquad (A24.40)$$

The integral equation (A24.5) is now modified as

$$\int d\Gamma_{1,\text{SL}} \left[g - \frac{D}{K(\mathbf{n})}h\right]\frac{\partial u}{\partial n_1} = \int d\Gamma_{1,\text{SL}} h u_0(\mathbf{r}_1). \qquad (A24.41)$$

The kinetic effect introduces only a small modification in the integral kernel as $g - Dh/K$, which is readily calculated. The growth rate v_n of the interface is obtained from the normal gradient $\partial u/\partial n$ as in Eq.(18.7). Therefore, a simple modification of the integral kernel G_{ij} can include the kinetic effect [50, 169].

In the directional solidification, the surface takes various periodic structures. In order to simulate an interface evolution with a periodic modulation with a periodicity λ, one imposes a periodic boundary condition $\zeta(x + \lambda) = \zeta(x)$ in x direction. For numerical simulation, one period of interface is decomposed into N mesh points. Periodic images of the interface are taken into account in the kernel as

$$\sum_{j=1}^{N} \tilde{G}_{ij} q_j = \sum_{j=1}^{N} \tilde{H}_{ij} u_j \qquad (A24.42)$$

with

$$\tilde{G}_{ij} = \sum_{m=-\infty}^{\infty} G_{i,j+mN}, \quad \text{and} \quad \tilde{H}_{ij} = \sum_{m=-\infty}^{\infty} H_{i,j+mN} \qquad (A24.43)$$

In practice, the influence of the m-th image far from the system in consideration are negligible. We usually take the effect of images within the range of $5l_\text{D}$ [167, 50]. For a dendrite in a channel with impermeable walls, the normal gradient of the diffusion field at the wall vanishes from the symmetry: $\partial u/\partial n = 0$. This means the existence of mirror images at both walls: $\zeta(-x) = \zeta(2\lambda - x) = \zeta(x)$ for a system within a channel of width λ. Then the system has a periodicity 2λ and one can appropriately modify Eq.(A24.42) in this case.

In the unidirectional solidification, the temperature gradient is applied on the system. If orientations of the temperature gradient and of the crystalline axis are different, the tilted structure appears [2, 66]. In this case the tip of the dendrite shifts transversally, and the locus of the tip is tilted from the temperature gradient by an angle ϕ. The system is invariant in a moving frame with a transversal velocity $v_x = v\tan\phi$ and a vertical pulling velocity $v_z = v$. The diffusion equation in quasi-stationary approximation reads as

$$\nabla^2 u + \frac{2}{l_D}\left(\frac{\partial u}{\partial z} + \tan\phi\frac{\partial u}{\partial x}\right) = 0. \qquad (A24.44)$$

The Green's function g and an integral kernel h are now modified to

$$g(\mathbf{r},\mathbf{r_1}) = \frac{1}{2\pi}\exp\left[-\frac{(\zeta - \zeta_1) - (x - x_1)\tan\phi}{l_D}\right] K_0\left[\frac{\sqrt{(x - x_1)^2 + (\zeta - \zeta_1)^2}}{l_D\cos\phi}\right] \qquad (A24.45)$$

and

$$h(\mathbf{r},\mathbf{r_1}) = \frac{1}{2\pi l_D}\exp\left[-\frac{(\zeta - \zeta_1) - (x - x_1)\tan\phi}{l_D}\right]\left[-(n_{1z} + n_{1x}\tan\phi)K_0\left(\frac{|\mathbf{r} - \mathbf{r_1}|}{l_D\cos\phi}\right)\right.$$
$$\left. - \frac{\mathbf{n_1}\cdot(\mathbf{r_1} - \mathbf{r})}{|\mathbf{r} - \mathbf{r_1}|\cos\phi}K_1\left(\frac{|\mathbf{r} - \mathbf{r_1}|}{l_D\cos\phi}\right)\right] - \frac{1}{2}\delta(\mathbf{r} - \mathbf{r_1}). \qquad (A24.46)$$

Further discretization and numerical simulation can be performed similarly as before [148].

A26 Eutectic Growth Theory by Jackson and Hunt

The surface supercooling ΔT at the eutectic crystal front is derived by Jackson and Hunt [91]. They considered both the lamellar and rod structure of an eutectic system, but here only the lamellar structure in two dimensions is summarized.

The eutectic crystal is growing from the liquid with a concentration c_∞. Concentrations of the crystal α and β phases are c_S^α and c_S^β, the periodicity of lamellae is λ, and the ratio of each phases are set η and $1 - \eta$ respectively. The material conservation requires the relation

$$c_\infty = c_S^\alpha \eta + c_S^\beta (1 - \eta), \qquad (A26.1)$$

and thus $\eta = (c_S^\beta - c_\infty)/\Delta c$ and $1 - \eta = (c_\infty - c_S^\alpha)/\Delta c$ with the miscibility gap $\Delta c = c_S^\beta - c_S^\alpha \ (> 0)$.

The dimensionless concentration field u defined by

$$u = \frac{c - c_E}{\Delta c} \qquad (A26.2)$$

satisfies the diffusion equation in the liquid

$$\frac{1}{D_c} \frac{\partial u}{\partial t} = \nabla^2 u + \frac{2}{l_D} \frac{\partial u}{\partial z} = 0. \qquad (A26.3)$$

The last equality holds in the quasi-stationary approximation. The diffusion length in the liquid is defined by means of the chemical diffusion constant D_c as

$$l_D = \frac{2D_c}{v}. \qquad (A26.4)$$

In solid the material diffusion is assumed negligible (one-sided model). The thermal diffusivity is so high in all the phases that the temperature gradient is constant as Eq.(26.1): $G_T = dT/dz = const$ or $T(z) = T_E + G_T z$.

The crystal-liquid interface is denoted as $z = \zeta(x,t)$. Then the material conservation at the boundary yields

$$-\left.\frac{\partial u}{\partial z}\right|_{z=\zeta} = \begin{cases} \dfrac{2}{l_D}\left[u(x,\zeta) - u_S^\alpha\right] & \text{for } 0 < x < \eta\lambda \text{ in } \alpha \text{ phase} \\[2ex] \dfrac{2}{l_D}\left[u(x,\zeta) - u_S^\beta\right] & \text{for } \eta\lambda < x < \lambda \text{ in } \beta \text{ phase}, \end{cases} \qquad (A26.5)$$

where $u_S^{\alpha,\beta} = (c_S^{\alpha,\beta} - c_E)/\Delta c$. The local equilibrium assumption at the interface is written as

$$\begin{aligned} \Delta T(x) &= T_E - T(\zeta) = -G_T \zeta \\[1ex] &= \begin{cases} m_\alpha \Delta c\, u(x,\zeta) + \dfrac{\gamma_{L\alpha} T_E}{L_\alpha}\kappa & \text{for } 0 < x < \eta\lambda \\[2ex] -m_\beta \Delta c\, u(x,\zeta) + \dfrac{\gamma_{L\beta} T_E}{L_\beta}\kappa & \text{for } \eta\lambda < x < \lambda. \end{cases} \end{aligned} \qquad (A26.6)$$

Here $m_{\alpha,\beta} = |dT_L/dc_L^{\alpha,\beta}|$ is the absolute slope of the liquidus lines with α and β crystal phases, κ is the curvatures, $\gamma_{L\alpha}$ and $\gamma_{L\beta}$ are the surface tensions between the liquid and α or β phase, L_α and L_β are the latent heat of α and β phases respectively.

From Eq.(A26.6) it is obvious that to obtain the supercooling ΔT at the average interface position $\langle \zeta \rangle$, we have to know the average concentration field $\langle u \rangle$ and the average curvature $\langle \kappa \rangle$.

The field u is expected to be modified periodically with the same periodicity λ of the interface modulation. Thus u can be expanded in a Fourier series as

$$u(x,z) = u_\infty + \sum_{n=-\infty}^{\infty} B_n e^{-\Lambda_n (z - \langle \zeta \rangle)} e^{iq_n x}, \tag{A26.7}$$

where $q_n = 2\pi n/\lambda$. Since u has to satisfy the diffusion equation in the quasi-stationary approximation (A26.3), the decay rate Λ_n in z direction is found to be

$$\Lambda_n = l_D^{-1} + \sqrt{l_D^{-2} + q_n^2}. \tag{A26.8}$$

The continuity equation (A26.5) is written at the average position $\langle \zeta \rangle$ as

$$\sum \Lambda_n B_n e^{iq_n x} = \frac{2}{l_D} \left[\Delta u(x) + \sum B_n e^{iq_n x} \right], \tag{A26.9}$$

where

$$\Delta u(x) = \begin{cases} 1 - \eta & \text{for } 0 < x < \eta\lambda \\ -\eta & \text{for } \eta\lambda < x < \lambda. \end{cases} \tag{A26.10}$$

Since the Fourier coefficient of $\Delta u(x)$ is calculated as

$$\frac{1}{\lambda} \int_0^\lambda dx \, \Delta u(x) e^{-iq_n x} = \frac{2}{\lambda q_n} e^{-i\lambda\eta q_n/2} \sin\left(\frac{\lambda\eta q_n}{2}\right), \tag{A26.11}$$

the coefficient B_n for $n \neq 0$ is determined as

$$B_n = \frac{4e^{-i\lambda\eta q_n/2} \sin(\lambda\eta q_n/2)}{l_D \lambda q_n (\Lambda_n - 2/l_D)}. \tag{A26.12}$$

Assuming the slow growth such that $\lambda \ll l_D$, the average concentration in front of the α or β phase is calculated as

$$\begin{aligned} \langle u^\alpha \rangle &= \frac{1}{\eta\lambda} \int_0^{\eta\lambda} u(x,\zeta) dx = u_\infty + B_0 + \frac{2\lambda}{l_D \eta} P(\eta) \\ \langle u^\beta \rangle &= \frac{1}{(1-\eta)\lambda} \int_{\eta\lambda}^\lambda u(x,\zeta) dx = u_\infty + B_0 - \frac{2\lambda}{l_D(1-\eta)} P(\eta), \end{aligned} \tag{A26.13}$$

where

$$P(\eta) = \sum_{n=1}^\infty \frac{\sin^2(\pi\eta n)}{(\pi n)^3}. \tag{A26.14}$$

From Eq.(A26.13) one obtains

$$\langle u^\alpha - u^\beta \rangle = \frac{c_L^\alpha - c_L^\beta}{\Delta c} = \frac{2\lambda}{l_D} \frac{P(\eta)}{\eta(1-\eta)} = \frac{\lambda v}{D_c} \frac{P(\eta)}{\eta(1-\eta)} \tag{A26.15}$$

or

$$v\Delta c = D_c \frac{c_L^\alpha - c_L^\beta}{\lambda/2} \frac{\eta(1-\eta)}{2P(\eta)}. \tag{A26.16}$$

The growth velocity Eq.(A26.16) corresponds to the phenomenological expressions (26.3) and (26.5) with Q_α in Eq.(26.5) at $c_\infty = c_E$ being given as

$$Q_\alpha = \frac{\eta(1-\eta)^2}{2P(\eta)}. \tag{A26.17}$$

In order to obtain the average curvature $\langle \kappa \rangle$, we use the relation of the force balance at the triple point, where three phase, liquid, α and β, meet. (See Fig.26.4). The surface tensions satisfy the relation (26.6a,26.6b) or:

$$\begin{aligned} \gamma_{L\alpha} \sin\theta_\alpha + \gamma_{L\beta} \sin\theta_\beta &= \gamma_{\alpha\beta} \\ \gamma_{L\alpha} \cos\theta_\alpha - \gamma_{L\beta} \cos\theta_\beta &= 0, \end{aligned} \tag{A26.18}$$

where the angles θ_α and θ_β are defined in Fig.26.4. These contact angles give the average curvature as

$$\begin{aligned} \langle \kappa^\alpha \rangle &= \frac{1}{\eta\lambda} \int_0^{\eta\lambda} \kappa dx = \frac{1}{\eta\lambda} \int_{-\theta_\alpha}^{\theta_\alpha} \frac{d\theta}{ds} ds \cos\theta = \frac{2}{\eta\lambda} \sin\theta_\alpha \\ \langle \kappa^\beta \rangle &= \frac{2}{(1-\eta)\lambda} \sin\theta_\beta. \end{aligned} \tag{A26.19}$$

Here we used the relation (A15.2) of the curvature κ, the angle θ of the normal vector and the arclength s: $\kappa = d\theta/ds$. These curvatures are the same with those obtained in Eq.(26.7) by assuming that the interface is a part of circle.

From Eq.(A26.6) the average supercooling at the interface, $z = \langle \zeta \rangle$ is expected to be constant for α and β phases as $\langle \Delta T^\alpha \rangle = \langle \Delta T^\beta \rangle = \Delta T$. Then the remaining unknown constant B_0 is determined as

$$\begin{aligned} u_\infty + B_0 &= -\frac{1}{m_\alpha + m_\beta} \left[\frac{2\lambda P(\eta)}{l_D} \left(\frac{m_\alpha}{\eta} - \frac{m_\beta}{1-\eta} \right) \right. \\ &\quad \left. + \frac{1}{\lambda\Delta c} \left(\frac{2\gamma_{L\alpha}T_E}{\eta L_\alpha} \sin\theta_\alpha + \frac{2\gamma_{L\beta}T_E}{(1-\eta)L_\beta} \sin\theta_\beta \right) \right]. \end{aligned} \tag{A26.20}$$

The interface supercooling ΔT is obtained as

$$\Delta T = \frac{1}{m_\alpha^{-1} + m_\beta^{-1}} \Delta c \frac{2\lambda}{l_D} \frac{P(\eta)}{\eta(1-\eta)}$$

$$+ \frac{2}{(m_\alpha^{-1} + m_\beta^{-1})\lambda} \left(\frac{\gamma_{L\alpha} T_E}{m_\alpha \eta L_\alpha} \sin\theta_\alpha + \frac{\gamma_{L\beta} T_E}{m_\beta(1-\eta)L_\beta} \sin\theta_\beta \right)$$
$$= \frac{1}{2}\Delta T_m \left(\frac{\lambda}{\lambda_m} + \frac{\lambda_m}{\lambda} \right). \tag{A26.21}$$

Here

$$\Delta T_m = \frac{4}{\eta(1-\eta)} \frac{\Delta c}{m_\alpha^{-1} + m_\beta^{-1}} \frac{P(\eta)}{l_D} \lambda_m \propto \sqrt{v} \tag{A26.22}$$

is the minimum value of the interface undercooling, and

$$\lambda_m^2 = \frac{l_D}{P(\eta)} [d_\alpha(1-\eta)\sin\theta_\alpha + d_\beta\eta\sin\theta_\beta] \propto \frac{1}{v} \tag{A26.23}$$

is the corresponding periodicity of the lamellar structure. The capillary lengths d_α and d_β are defined as

$$d_\alpha = \frac{\gamma_{L\alpha} T_E}{m_\alpha L_\alpha \Delta c} \qquad \text{and} \qquad d_\beta = \frac{\gamma_{L\beta} T_E}{m_\beta L_\beta \Delta c}. \tag{A26.24}$$

These equations (A26.21-A26.24) are more general than Eq.(26.10).

References

[1] M.Abramowitz and I.A.Stegun, ed. 'Handbook of mathematical functions', (Dover, New York, 1972).

[2] S.Akamatsu, G.Faivre and T.Ihle, 'Symmetry-broken double fingers and sea-weed patterns in thin-film directional solidification of a nonfaceted cubic crystal', Phys. Rev. E**51**, 4751-4773 (1995).

[3] Y.Akutsu, N.Akutsu and T.Yamamoto, 'Universal jump of Gaussian curvature of the facet edge of a crystal', Phys. Rev. Lett. **61**, 424-427 (1988), ibd. **62**, 2637 (1989).

[4] C.Alfonso, J.M.Bermond, J.C.Heyraud and J.M.Métois 'The meandering of steps and the terrace width distribution on clean Si(111)', Surf. Sci. **262**, 371-381 (1992).

[5] R.Almgren, W.-S.Dai and V.Hakim, 'Scaling behavior in anisotropic Hele-Shaw flows', Phys. Rev. Lett. **71**, 3461-3464 (1993).

[6] A.F. Andreev, 'Faceting phase transitions of crystals', Zh.Exsp.Teor.Fiz. **80**, 2042-2052 (1981). [Sov. Phys. JETP **53**, 1063-1069 (1981).]

[7] J.E.Avron, H.van Beijeren, L.S.Schulman and R.K.P.Zia, 'Roughening transition, surface tension and equilibrium droplet shapes in a two-dimensional Ising system', J. Phys. A**15**, L81-L86 (1982).

[8] G.S.Bales and A.Zangwill, 'Morphological instability of a terrace edge during step-flow growth', Phys. Rev. B**41**, 5500-5508 (1990).

[9] S.Balibar, F.Gallet and R.Rolley, 'The dynamic roughening of crystals', J. Cryst. Growth **99**, 46-53 (1990).

[10] A.-L.Barabási and H.E.Stanley, 'Fractal concepts in surface growth', (Camridge, New York, 1995).

[11] M.N.Barber, A.Barbieri and J.S. Langer, 'Dynamics of dendritic sidebranching in the two-dimensional symmetric model of solidification', Phys. Rev. A**36**, 3340-3349 (1987).

[12] R.Becker and W.Döring, 'Kinetische Behandlung der Keimbildung in übersättigten Dämpfen', Ann. Phys. **24**, 719-752 (1935).

[13] I.Bena, C.Misbah and A.Valance, 'Nonlinear evolution of a terrace edge during step flow growth', Phys. Rev. B**47**, 7408-7419 (1993).

[14] M. Ben Amar and E. Brener, 'Theory of pattern selection in three-dimensional nonaxisymmetric dendritic growth', Phys. Rev. Lett. **71**, 589-592 (1993).

[15] M.Ben Amar and E.Brener, 'Parity-broken dendrites', Phys. Rev. Lett. **75**, 561-564 (1995).

[16] C.M.Bender and S.A. Orszag, 'Advanced mathematical methods for scientists and engineers', (Mc-Graw-Hill, New York, 1978).

[17] E. Ben-Jacob, N.D. Goldenfeld, J.S. Langer and G. Schön, 'Dynamics of interfacial pattern formations', Phys. Rev. Lett. **51**, 1930-1932 (1983), and 'Boundary-layer model of pattern formation in solidification', Phys. Rev. A**29**, 330-334 (1984).

[18] E. Ben-Jacob, N. Goldenfeld, B.G. Kotliar, and J.S. Langer, 'Pattern selection in dendritic solidification', Phys. Rev. Lett. **53**, 2110-2113 (1984).

[19] E. Ben-Jacob, R. Godbey, N.D. Goldenfeld, J. Koplik, H.Levine, T. Mueller, and L.M. Sander, 'Experimental demonstration of the role of anisotropy in interfacial pattrn formation', Phys. Rev. Lett. **55**, 1315-1318 (1985).

[20] E. Ben-Jacob, G. Deutscher, P. Garik, N.D. Goldenfeld, and Y. Lereath, 'Formation of a dense branching morphology in interfacial growth', Phys. Rev. Lett. **57**, 1903-1906 (1986).

[21] E. Ben-Jacob, P. Garik, and D. Grier, 'Interfacial pattern formation far from equilibrium', Superlatt. Microstruct. **3**, 599-615 (1987).

[22] E. Ben-Jacob, A. Tenenbaum, O. Shochet, O. Avidan, 'Holotransformations of bacterial colonies and genome cybernetics', Physica **A 202**, 1-47 (1994).

[23] P.Bennema and G.H.Gilmer, 'Kinetics of crystal growth', in *Crsytal growth: an Introduction*, ed. P.Hartman, (North-Holland, Amsterdam, 1973) p.263-327.

[24] W.F. Berg, 'Crystal growth from solutions', Proc. Roy. Soc. A**164**, 79-95 (1938).

[25] K. Binder, ed., 'Monte Carlo Methods in Statistical Physics', (Springer, Berlin, 1979).

[26] U.Bisang and J.H.Bilgram, 'Shape of the tip and the formation of sidebranches of Xenon dendrites', Phys. Rev. Lett. **75**, 3898-3901 (1995).

[27] A.G.Borisov, O.P.Fedorov and V.V.Maslov, 'Growth of succinonitrile dendrites in different crystallographic directions', J. Crystal Growth **112**, 463-466 (1991).

[28] P.Bouissou, A.Chiffaudel, B. Perrin and P.Tabeling, 'Dendritic side-branching forced by an external flow', Europhys. Lett. **13**, 89-94 (1990).

[29] K.Brattkus and C. Misbah, 'Phase dynamics in directional solidification', Phys. Rev. Lett. **64**, 1935-1938 (1990).

[30] C.A.Brebbia, 'The boundary element method for engineers', (Pentech, London, 1978).

[31] E.A.Brener, M.B.Geilikman and D.E.Temkin, 'Growth of a needle-shaped crystal in a channel', Zh.Eksp.Teor. Fiz. **94**,241-255 (1988) [Sov. Phys. JETP **67**, 1002-1009 (1988).]

[32] E.A.Brener, 'Influence of kinetic effects on the growth of a two-dimensional dendrite', Zh.Exsp.Teor.Fiz. **96**, 237-245 (1989). [Sov. Phys. JETP **69**, 133-137 (1989).

[33] E.Brener and V.I.Mel'nikov, 'Pattern selection in two-dimensional dendritic growth', Adv. Phys. **40**, 53-97 (1991).

[34] E.Brener, H.Müller-Krumbhaar and D.Temkin, 'Kinetic phase diagram and scaling relations for stationary diffusional growth', Europhys. Lett. **17**, 535-540 (1992).

[35] E.Brener, H.Müller-Krumbhaar, Y.Saito, and D.Temkin, 'Crystal growth in a channel: Numerical study of the one-sided model', Phys. Rev. E**47**, 1151-1155 (1993).

[36] E. Brener, 'Needle-crystal solution in three-dimensional dendritic growth', Phys. Rev. Lett. **71** 3653-3656 (1993).

[37] E.Brener, T.Ihle, H.Müller-Krumbhaar, Y.Saito and K.Shiraishi, 'Fluctuation effects on dendritic growth', Physica A **204**, 96-110 (1994).

[38] E.Brener and D.Temkin, 'Noise-induced sidebranching in the three-dimensional nonaxisymmetric dendritic growth', Phys. Rev. E**51**, 351-359 (1995).

[39] J.Q.Broughton, G.H.Gilmer and K.A.Jackson, 'Crystallization rates of a Lennard-Jones liquid', Phys. Rev. Lett. **49**, 1496-1500 (1982).

[40] R. Brower, D. Kessler, J. Koplik and H. Levine, 'Geometrical approach to moving-interface dynamics', Phys. Rev. Lett. **51**, 1111-1114 (1983); Phys. Rev. A**29**, 1335 (1984).

[41] H. Brune, C. Romalnczyk, H. Röder and K. Kern, 'Mechanism of the transition from fractal to dendritic growth of surface aggregates', Nature **369**, 469-471 (1994).

[42] C.W.Bunn, 'Crystal growth from solution II. Concentration gradients and the rates of growth of crystals', Disc. Farady Soc. **5**, 132-144 (1949).

[43] W.K.Burton, N.Cabrera and F.C.Frank, 'The growth of crystals and the equilibrium structure of their surfaces', Phil. Trans. Royal Soc. London, **243**, 299-358 (1951).

[44] N. Cabrera and M.M. Levine, 'On the dislocation theory of evaporation of crystals', Phil. Mag. **1**, 450-458 (1956).

[45] S.-K.Chan, H.-H.Reimer and M.Kahlweit, 'On the stationary growth shapes of NH$_4$Cl dendrites', J. Cryst. Growth **32**, 303-315 (1976).

[46] A.A.Chernov, 'The kinetics of the growth forms of crystals', Kristallografiya, **7**, 895-898 (1962), [Sov.Phys. Crystallogr. **7** 728-730 (1963)].

[47] A.A.Chernov, 'Stability of faceted shapes', J. Cryst. Growth **24/25**, 11-31 (1974).

[48] A.A. Chernov, 'Crystallization Processes', in *Modern Crystallography III*, ed. A.A. Chernov, (Springer, Berlin, 1984) p.1-297.

[49] S.T.Chui and J.D.Weeks, 'Phase transition in the two-dimensional Coulomb gas, and the interfacial roughening transition', Phys. Rev. B14, 4978-4982 (1976).

[50] A.Classen, C.Misbah, H.Müller-Krumbhaar and Y.Saito, 'Directional solidification with interface dissipation', Phys. Rev. A43, 6920-6933 (1991).

[51] S.R.Coriell and R.F. Sekerka, 'The effect of the anisotropy of surface tension and interface kinetics on morphological stability', J. Cryst. Growth 34, 157-163 (1976).

[52] P.Coullet, R.E.Goldstein and G.H.Gunaratne, 'Parity-breaking transition of modulated patterns in hydrodynamic system', Phys. Rev. Lett. 63 1954-1957 (1989).

[53] V.Datye and J.S.Langer, 'Stability of thin lamellar eutectic growth', Phys. Rev. B24, 4155-4169 (1981).

[54] A.Dougherty, P.D.Kaplan and J.D.Goluub, 'Development of side branching in dendritic crystal growth', Phys. Rev. Lett. 58, 1652-1655 (1987).

[55] W.Eckhaus, 'Studies on Nonlinear Stability Theory', Springer tracts in natural philosophy, Vol.6, (Springer, Berlin, 1965).

[56] A. Einstein, 'On the movement of small particles suspended in stationary liquids demanded by the molecular kinetics theory of heat', Ann. Phys. (Leipzig) 17, 549 (1905), [in *Investigation on the Theory of Brownian Movement*, (Dover, New York, 1956) p.1-18.]

[57] G.Faivre and J.Mergy, 'Tilt bifurcation and dynamical selection by tilt domains in thin-film lamellar eutectic growth: Experimental evidence of a tilt bifurcation', Phys. Rev. A45, 7320-7329 (1992).

[58] G.Faivre and J.Mergy, 'Dynamical wavelength selection by tilt domains in thin-film lamellar eutectic growth', Phys. Rev. A46, 963-972 (1992).

[59] R.P.Feynman, 'Statistical Mechanics', (Benjamin, Reading, 1972).

[60] M.P.A.Fisher, D.S.Fisher and J.D.Weeks, 'Agreement of capillary-wave theory with exact results for the interface profile of the two-dimensional Ising model', Phys. Rev. Lett. 48, 368 (1982).

[61] D.S.Fisher and J.D. Weeks, 'Shape of crystals at low temperatures: Absence of quantum roughening', Phys. Rev. Lett. 50, 1077-1080 (1983).

[62] F.C. Frank, 'The influence of dislocation on crystal growth', Disc. Faraday Soc. 5, 48-54 (1949).

[63] J.Frenkel, Phys. Z. Sowjetunion 1, 498 (1932).

[64] H.Fujikawa and M.Matsushita, 'Fractal growth of Bacillus subtilis on agar plates', J. Phys. Soc. Jpn. **58**, 3875-3878 (1989); 'Bacterial fractal growth in the concentration field of nutrient', J. Phys. Soc. Jpn. **60**, 88 (1991).

[65] Y. Furukawa and W. Shimada, 'Experimental study of the pattern formation of ice crystal grown in supercooled water', in *Pattern Formation in complex dissipative systems*, ed. S.Kai, (World Scientific, Singapore, 1992) p.14-22.

[66] Y. Furukawa and W. Shimada, 'Three-dimensional pattern formation during growth of ice dendrites — Its relation to universal law of dendritic growth', J. Cryst. Growth **128**, 234-239 (1993).

[67] M.Fusimi, 'Ransu (Random number)', (Univ. of Tokyo Publ., Tokyo, 1989) (in japanese).

[68] F.Gallet, S.Balibar and E.Rolley, 'The roughening transition of crystal surfaces. II. Experiments on static and dynamic properties near the first roughening transition of hcp ^{4}He', J. Phys. **48**, 369-377 (1987).

[69] G.H.Gilmer and P.Bennema, 'Simulation of crystal growth with surface diffusion', J. Appl. Phys. **43**, 1347-1360 (1972).

[70] G.H.Gilmer and K.A.Jackson, 'Computer simulation of crystal growth', in *Crystal growth and materials*, ed. E.Kaldis and H.J.Scheel, (North Holland, Amsterdam, 1977) p.80-114.

[71] M.E.Glicksman and N.B.Singh, 'Effect of crystal-melt interfacial energy anisotropy on dendritic morphology and growth kinetics', J. Cryst. Growth **98**, 277-284 (1989).

[72] E.E. Gruber and W.W. Mullins, 'On theory of anisotropy of crystalline surface tension', J. Phys. Chem. Solids **28**, 875 (1967).

[73] R.N.Grugel and Y.Zhou, 'Primary dendrite spacing and the effect of off-axis heat flow', Metall. Trans. **20A**, 969-973 (1989).

[74] P.Hartman, ed. 'Crystal growth: an introduction', (North-Holland, Amsterdam, 1973).

[75] J.S.S. Hele-Shaw, Nature **58**, 34-36 (1898).

[76] D.W.Heermann, 'Computer simulation methods in theoretical physics', (Springer, Berlin, 1986).

[77] H. Hertz, Ann. Phys. Lpz. **17**, 193 (1882).

[78] J.C.Heyraud and J.J.Métois, 'Equilibrium shape and temperature; Lead on graphite', Surf. Sci. **128**, 334-342 (1983).

[79] J.C.Heyraud and J.J.Metois, 'Equilibrium shape of an ionic crystal in equilibrium with its vapour (NaCl)', J. Cryst. Growth **84**, 503-508 (1987).

[80] W.B.Hillig, 'A derivation of classical two-dimensional nucleation kinetics and the associated crystal growth laws', Acta Metall. **14**, 1868-1869 (1966).

[81] J.D.Hoffman, 'Regime III crystallization in melt-crystallized polymers: The variable cluster model of chain folding', Polymer **24**, 3-26 (1983).

[82] H. Honjo, S. Ohta and M. Matsushita, 'Irregular fractal-like crystal growth of Ammonium Chloride', J. Phys. Soc. Jpn. **55**, 2487-2490 (1986).

[83] G.Horvay and J.W.Cahn, 'Dendritic and spheroidal growth', Acta Metall. **9**, 695-705 (1961).

[84] S-C. Huang and M.E. Glicksman, 'Fundamentals of dendritic solidification, I: Steady-state tip growth', and 'II: Developement of sidebranch structure', Acta Metall. **29**, 701-716 and 717-734 (1981).

[85] D.T.J. Hurle, ed. 'Handbook of crystal growth', Vol.1,2 and 3, (North-Holland, Amsterdam, 1993).

[86] R.Q.Hwang, J.Schröder, G.Gunther and R.J.Behm, 'Fractal growth of two-dimensional islands: Au on Ru(0001)', Phys. Rev. Lett. **67** 3279-3282 (1991).

[87] T.Ihle and H.Müller-Krumbhhar, 'Fractal and compact growth morphologies in phase-transitions with diffusion-transport', Phys. Rev. **E 49**, 2972 (1994).

[88] T.Irisawa, Y.Arima and T.Kuroda, 'Periodic changes in the structure of a surface growing under MBE conditions', J. Cryst. Growth **99**, 491-495 (1990).

[89] G.P.Ivantsov, Dokl. Akad. Nauk USSR **58** (1947) 1113.

[90] K.A.Jackson, 'Mechanism of growth', in *Liquid metals and solidification*, (Am. Soc. Metals, Cleavland, Ohio,1958) p.174-186.

[91] K.A.Jackson and J.D.Hunt, 'Lamellar and rod eutectic growth', Trans. Metall. Soc. AIME **236**, 1129-1142 (1966).

[92] C.Jayaprakash, W.F.Saam and S.Teitel, 'Roughening and facet formation in crystals', Phys. Rev. Lett. **50**, 2017-2020 (1983).

[93] C.Jayaprakash and W.F.Saam, 'Thermal evolution of crystal shapes: The fcc crystal', Phys. Rev. **B30**, 3916-3928 (1983).

[94] A.Karma, 'Beyond steady-state lamellar eutectic growth', Phys.Rev. Lett. **59**, 71-74 (1987).

[95] K.Kassner and C.Misbah, 'Parity breaking in eutectic growth', Phys. Rev. Lett. **65**, 1458-1461 (1990). Errata, Phys. Rev. Lett. **66**, 522 (1991).

[96] K.Kassner and C.Misbah, 'Growth of lamellar eutectic structures: The axisymmetric state', Phys. Rev. **A44**, 6513-6532 (1991).

[97] K.Kassner and C.Misbah, 'Spontaneous parity-breaking transition in directional growth of lamellar eutectic structure', Phys. Rev. **A44**, 6533-6543 (1991).

[98] K.Kassner and C.Misbah, 'Similarity laws in eutectic growth', Phys. Rev. Lett. **66**, 445-448 (1991).

[99] K.Kassner, 'Morphological instability: dendrites, seaweed and fractals', in *Science and technology of crystal growth*, eds. J.P.van der Eerden and O.S.L. Bruinsma, (Kluwer Academic, Dordrecht, 1995) p.193-208.

[100] D.A. Kessler, J. Koplik and H. Levine, 'Geometrical models of interface evolution. II. Numerical simulation', Phys. Rev. **A30**, 3161-3174 (1984).

[101] D.A. Kessler, J. Koplik and H. Levine, 'Geometrical models of interface evolution. III. Theory of dendritic growth', Phys. Rev. **A31**, 1712-1717 (1985).

[102] D.A.Kessler, J.Koplik and H.Levine, 'Steady-state dendritic crystal growth', Phys. Rev. **A33**, 3352-3357 (1986).

[103] D.A.Kessler and H.Levine, 'Velocity selection in dendritic growth', Phys. Rev. **B33**, 7867-7870 (1986).

[104] D.A.Kessler, J.Koplik and H.Levine, 'Dendritic growth in a channel', Phys. Rev. **A34**, 4980-4987 (1986).

[105] D.A. Kessler, J. Koplik and H. Levine, 'Pattern selection in fingered growth phenomena', Adv. Phys. **37**, 255-339 (1988).

[106] D.A. Kessler and H. Levine, 'Steady-state cellular growth during directional solidification', Phys. Rev. **A39**, 3041-3052 (1989).

[107] D.A. Kessler and H. Levine, 'Computational approach to steady-state eutectic growth', J.Cryst. Growth **94**, 871-879 (1989).

[108] J.S.Kirkaldy, 'Predicting the pattern in lamellar growth', Phys. Rev. **B30**, 6889-6895 (1984).

[109] R.Kobayashi, 'Simulations of three dimensional dendrites', in *Pattern formation in complex dissipative systems*, ed. S. Kai, (World Scientific, Singapore, 1992) p.121-128.

[110] R.Kobayashi, 'Modeling and numerical simulations of dendritic crystal growth', Physica **D63**, 410-423 (1993).

[111] R.Kobayashi, 'A numerical approach to three-dimensional dendritic solidification', Exp. Math. **3**, 59-81 (1994).

[112] W.Kossel, 'Zur Theorie des Kristallwachstums', Nachr. Ges. Wiss. Göttingen (1927) p.135-143.

[113] J.M.Kosterlitz and D.J. Thouless, 'Ordering, metastability and phase transitions in two-dimensional systems', J. Phys. **C6**, 1181-1203 (1973).

[114] J.M. Kosterlitz, 'The critical properties of the two-dimensional xy model', J. Phys. **C7**, 1046-1060 (1974).

[115] J.M. Kosterlitz, 'The d-dimensional Coulomb gas and the roughening transition', J. Phys. **C10**, 3753-3760 (1977).

[116] M.Knudsen, Ann. Phys. Lpz. **47**, 697 (1915).

[117] R.Kupferman, O.Shochet, E.Ben-Jacob and Z.Schuss, 'Phase-field model: Boundary layer, velocity of propagation and the stability spectrum', Phys. Rev. B**46**, 16045-16057 (1992).

[118] Y.Kuramoto and T.Tsuzuki, 'Persistent propagation of concentration waves in dissipative media far from thermal equilibrium', Prog. Theor. Phys. **55**, 356-369(1976).

[119] T.Kuroda, T.Irisawa and A.Ookawa, 'Growth of a poyhedral crystal from solution and its morphological stability', J. Cryst. Growth **42** 41-46 (1977).

[120] W. Kurz, D. Fisher, 'Fundamentals of Solidification', (Trans Tech Publ., Aedermannsdorf, Switzerland 1989).

[121] L.D.Landau and E.M.Lifshitz, 'Statistical Physics', (Pergamon, Oxford, 1963).

[122] J.S. Langer, 'Instabilities and pattern formation in crystal growth', Rev. Mod. Phys. **52**, 1-28 (1980)

[123] J.S.Langer, 'Dendritic sidebranching in the three-dimensional symmetric model in the presence of noise', Phys. Rev. A**36**, 3350-3358 (1987).

[124] J.S.Langer, 'Lectures in the theory of pattern formation', in *Chance and Matter*, ed. J. Souletie, J.Vannimenus and R.Stora (North Holland, Amsterdam, 1987) p.629-711.

[125] A.V.Latyshev, H.Minoda, Y.Tanishiro and K.Yagi, 'Ultra High Vacuum Reflection Electron Microscopy study of step-dislocation interaction on Si(111) surface', Jpn. J. Appl. Phys. **34**, 5768-5773 (1995).

[126] A.V.Latyshev, H.Minoda, Y.Tanishiro and K.Yagi, 'Dynamical step edge stiffness on the Si (111) surface', Jpn. J. Appl. Phys. (1995) to appear.

[127] H.J.Leamy, G.H.Gilmer and K.A.Jackson, 'Statistical thermodynamics of clean surfaces', in *Surface Physics of Materials*, Vol.1, ed. J.M. Blakely, (Academic, New York, 1975) p.121-188.

[128] T.D.Lee and C.N. Yang, 'Statistical theory of equations of state and phase transitions. II. Lattice gas and Ising model', Phys. Rev. **87**, 410-419 (1952).

[129] E.H.Lieb and F.Y.Wu, 'Two-dimensional ferroelectric models', in *Phase transitions and critical phenomena*, ed. C.Domb and M.S. Green, (Academic, London, 1972) Vol.1, p.331-490.

[130] M. Matsushita, Y.Hayakawa and Y.Sawada, 'Fractal structure and cluster statistics of zinc-metal trees deposited on a line electrode', Phys. Rev. A**32**, 3814-3816 (1985).

[131] M.Matsushita and H.Yamada, 'Dendritic growth of single viscous finger under the influence of linear anisotropy', J. Cryst. Growth **99**, 161-164 (1990).

[132] M. Matsushita, H. Fujikawa, 'Diffusion-limited growth in bacterial colony formation', Physica **A 168**, 498-506 (1990).

[133] O. Martin, N.D. Goldenfeld, 'Origin of sidebranching in dendritic growth', Phys. Rev. A**35**, 1382-1390 (1987).

[134] P.Meakin, 'Diffusion-controlled cluster formation in 2-6-dimensional space', Phys. Rev. A**26**, 1495-1507 (1983).

[135] D.I.Meiron, 'Selection of steady states in the two-dimensional symmetric model of dendritic growth', Phys. Rev. A**33**, 2704-2715 (1986).

[136] J.J.Métois and J.C.Heyraud, 'Analysis of the critical behaviour of curved regions in equilibrium shapes of In crystals', Surf.Sci. **180**, 647-653 (1987).

[137] L.V.Mikheev and A.A. Chernov, 'Mobility of a diffuse simple crystal-melt interfaces', J.Cryst. Growth **112**, 591-596 (1991).

[138] C.Misbah, 'Velocity selection for needle crystals in the 2-D one-sided model', J. Phys. **48**, 1265-1272 (1987).

[139] S. Miyashita, M. Uwaha and Y. Saito, 'Experimental evidence of a dynamical scaling in a fractal aggregation growth', (private communication, 1996).

[140] W.W. Mullins and R.F. Sekerka, 'Morphological stability of a particle growing by diffusion or heat flow', J. Appl. Phys. **34**, 323-329 (1963).

[141] W.W. Mullins and R.F. Sekerka, 'Stability of a planar interface during solidification of a dilute binary alloy', J. Appl. Phys. **35**, 444-451 (1964).

[142] H.Müller-Krumbhaar, T.W.Burkhardt and D.M.Kroll, 'A generalized kinetic equation for crystal growth', J. Cryst. Growth **38**, 13-22 (1977).

[143] H.Müller-Krumbhaar and W. Kurz, 'Solidification', in *Phase transformations in materials*, ed. P.Haasen, (VCH, Weinheim, 1991) p.553-632.

[144] K.Nagashima and Y.Furukawa, 'Nonequilibrium effect of anisotropic interface kinetics for the directional growth of ice crystal', J. Cryst. Growth (1995) submitted.

[145] P.Nozières and F.Gallet, 'The roughening transition of crystal surfaces. I. Static and dynamic renormalization theory, crystal shape and facet growth', J. Phys. **48**, 353-367 (1987).

[146] P.Nozières, 'Shape and growth of crystals', in *Solids far from equilibrium*, ed. C.Godrèche, (Cambridge, Cambridge, 1991) p.1-154.

[147] T.Ohachi and I.Taniguchi, 'Roughening transition for the ionic-electronic mixed superionic conductor $\alpha - Ag_2S$', J. Cryst. Growth **65**, 84-88 (1983).

[148] T. Okada and Y.Saito, 'Simulation of unidirectional solidification with a tilted crystalline axis', Phys. Rev. E (1996) to appear.

[149] L.Onsager, 'Crystal statistics I. A two-dimensional model with an order-disorder transitions', Phys. Rev. **65**, 117-149 (1944).

[150] P.Ossadnik, 'Multiscaling analysis of large-scale off-lattice DLA', Physica A **176**, 454-462 (1991).

[151] P.Oswald, J.Malthête and P.Pelcé, 'Free growth of a thermotropic columnar mesophase: supersaturation effects', J.Phys. France **50**, 2121-2138 (1989).

[152] L.J. Paterson, 'Radial fingering in a Hele Shaw cell', J. Fluid Mech. **113**, 513-529 (1981).

[153] L.Pauling, 'The structure and entropy of ice and of other crystals with some randomness of atomic arrangement', J. Am. Chem. Soc. **57**, 2680-2684 (1935).

[154] P.Pelcé and Y.Pomeau, 'Dendrite in the small undercooling limit', Studies in Appl. Math. **74**, 245-258 (1986).

[155] R. Pieters and J.S. Langer, 'Noise-driven sidebranching in the boundary-layer model of dendritic solidification', Phys. Rev. Lett. **56**, 1948-1951 (1986).

[156] Y.Pomeau and M.Ben Amar, 'Dendritic growth and related topics', in *Solids far from equilibrium*, ed. C.Godrèche, (Cambridge, Cambridge, 1991) p.365-431.

[157] X.W. Qian and H.Z. Cummins, 'Dendritic sidebranching initiation by a localized heat pulse', Phys. Rev. Lett. **64**, 3038-3041 (1990).

[158] E.Rolley, C.Guthmann, E.Chevalier and S. Balibar, 'The static and dynamic properties of vicinal surfaces on Helium 4 crystals', J. Low Temp. Phys. **99**, 851-886 (1995).

[159] C. Rottman and M. Wortis, 'Exact equilibrium crystal shapes at nonzero temperature in two dimensions', Phys. Rev. **B24**, 6274-6277 (1981).

[160] C.Rottman, M.Wortis, J.C.Heyraud and J.J.Métois, 'Equilibrium shapes of small lead crystal: Observation of Pokrovsky-Talapov critical behavior', Phys. Rev. Lett. **52** 1009-1012 (1984).

[161] F.Saam, 'Comment on "Universal jump of Gaussian curvature at the facet edge of a crystal"', Phys. Rev. Lett. **62**, 2636 (1989).

[162] P.G.Saffman and G. Taylor, 'The penetration of a fluid into a porous medium or Hele-Shaw cell containing a more viscous liquid', Proc. R. Soc. London, Ser.A **245**, 312-329 (1958).

[163] Y.Saito, 'Self-consistent calculation of statics and dynamics of the roughening transition', Z. Phys. B **32**, 75-82 (1978).

[164] Y.Saito and H.Müller-Krumbhaar, 'Two-dimensional Coulomb gas: A Monte Carlo study', Phys. Rev. **B23**, 308-315 (1981).

[165] Y.Saito, G.Goldbeck-Wood and H.Müller-Krumbhaar, 'Dendritic crystallization: Numerical study of the one-sided model', Phys. Rev. Lett. **58**, 1541-1543 (1987); 'Numerical simulation of dendritic growth', Phys. Rev. **A38**, 2148-2157 (1988).

[166] Y.Saito and T.Ueta, 'Monte Carlo studies of equilibrium and growth shapes of a crystal', Phys. Rev. **A40**, 3408-3419 (1989).

[167] Y.Saito, C.Misbah and H.Müller-Krumbhaar, 'Directional solidification: Transition from cell to dendrite', Phys. Rev. Lett. **63**, 2377-2380 (1989).

[168] Y. Saito, C. Misbah, H. Müller-Krumbhaar, T. Ueta and M. Uwaha, 'Formation of Growth Patterns in a Diffusion Field', in *Formations, Dynamics and Statistics of Patterns,* ed. K. Kawasaki, M. Suzuki and A. Onuki, (World Scientific, Singapore, 1990) p.236-284.

[169] Y.Saito and T.Sakiyama, 'Kinetic effect on 2D dendritic growth', J. Cryst. Growth **128**, 224-228 (1993).

[170] Y. Saito and M. Uwaha, 'Fluctuation and instability of steps in a diffusion field', Phys. Rev. B**49**, 10677-10692 (1994).

[171] L.I.Schiff, 'Quantum mechanics', (McGraw-hill, Auckland, 1968).

[172] R.L.Schwoebel and E.J.Shipsey, 'Step motion on crystal surfaces', J. Appl. Phys. **37**, 3682-3686 (1966).

[173] A.Seeger, 'Diffusion problems associated with the growth of crystals from dilute solution', Phil. Mag. **44**, 1-7 (1953).

[174] V.Seetharaman and R.Trivedi, 'Eutectic growth: Selection of interlamellar spacings', Metall. Trans. **19A**, 2955-2964 (1988).

[175] O.Shochet, K.Kassner, E.Ben-Jacob, S.G.Lipson and H.Müller-Krumbhaar, 'Morphology transitions during non-equilibrium growth II', Physica A**187**, 87-111 (1992).

[176] O.Shochet and E.Ben-Jacob, 'Coexistence of morphologies in diffusive patterning', Phys. Rev. E**48**, R4168-R4171 (1993).

[177] G.I.Sivashinsky, Acta Astraunautica **4**, 1177- (1977).

[178] B.S.Swartzentruger, Y-W.Mo, R.Kariotis, M.G.Lagally and M.B.Webb, 'Direct determination of step and kink energies on vicinal Si(001)', Phys. Rev. Lett. **65**, 1913-1916 (1990).

[179] A.Toda, 'Growth kinetics of polyethylene single crystals', in *Crystallization of polymers*, ed. M.Dosière, (Kluwer Academic, Amsterdam, 1993) p.141-152.

[180] R. Trivedi, 'Interdendritic Spacing: Part II. A Comparison of theory and experiment', Metall. Trans. **15A**, 977-982 (1984).

[181] R. Trivedi and K.Somboonsuk, 'Pattern formation during the directional solidification of binary systems', Acta Metall. **33**, 1061-1068 (1985).

[182] R.Trivedi, V.Seetharaman and M.A.Eshelman, 'The effect of interface kinetics anisotropy on the growth direction of cellular microstructures', Metall. Trans. **22A**, 585-593 (1991).

[183] L.H.Ungar and R.A.Brown, 'Cellular interface morphologies in directional solidification. I. The one-sided model' Phys. Rev. B**29**, 1367-1380 (1984); 'II. The

effect of grain boundaries', Phys. Rev.B**30**, 3993-3999 (1984); 'III. The effect of heat transfer and solid diffusivity', Phys. Rev. B**31**, 5923-5930 (1985); 'IV. The formation of deep cells', Phys. Rev. B**31**, 5931-5940 (1985).

[184] M.Uwaha, 'Asymptotic growth shapes developed from two-dimensional nuclei', J. Cryst. Growth **80**, 84-90 (1987).

[185] M.Uwaha and Y.Saito, 'Fractal-to-compact transition and velocity selection in aggregation from lattice gas', J. Phys. Soc. Jpn. **57**, 3285-3288 (1988); 'Aggregation growth in a gas of finite density: Velocity selection via fractal dimension of diffusion-limited aggregation', Phys. Rev. A**40**, 4716-4723 (1989).

[186] M.Uwaha and Y.Saito, 'Kinetic smoothing and roughening of a step with surface diffusion', Phys. Rev. Lett. **68**, 224-227 (1992).

[187] H.van Beijeren, 'Exactly solvable model for the roughening transition of a crystal surface', Phys. Rev. Lett. **38**, 993-996 (1977).

[188] J.P.van der Eerden, D. Kaschchiev and P.Bennema, 'Surface migration of small crystallites: A Monte Carlo simulation with continuous time', J.Cryst. Growth **42**, 31-34 (1977).

[189] T.Vicsek, 'Fractal growth phenomena', (World Scientific, Singapore, 1989).

[190] J.Villain, 'Continuum models of crystal growth from atomic beams with and without desorption', J. Phys. I **1**, 19-42 (1991).

[191] R.F.Voss, 'Multiparticle fractal aggregation', J. Stat. Phys. **36**, 861-872 (1984).

[192] X.-S.Wang, J.L.Goldber, N.C.Bartelt, T.L.Einstein and E.D.Williams, 'Terrace-width distributions on vicinal Si(111)', Phys. Rev. Lett. **65**, 2430-2433 (1990).

[193] G.H. Wannier, 'The statistical problem in cooperative phenomena', Rev. Mod. Phys. **17**, 50-60 (1945).

[194] J.D. Weeks and G.H.Gilmer, 'Dynamics of crystal growth', Adv. Chem. Phys. **40**, 157-228 (1979).

[195] J.D. Weeks, 'The roughening transition', in *Ordering in Strongly Fluctuating Condensed Matter Systems*, ed. T.Riste(New York, Plenum, 1980) p.293-317.

[196] H.A. Wilson, 'On the velocity of solidification and viscosity of supercooled liquids', Philos. Mag. **50**, 238-250 (1900).

[197] T.A.Witten and L.M.Sanders, 'Diffusion-limited aggregation, a kinetic critical phenomena', Phys. Rev. Lett. **47** 1400-1403 (1981); 'Diffusion limited aggregation', Phys. Rev. B**27**, 5686 (1983).

[198] P.E.Wolf, F.Gallet, S.Balibar, E.Rolley and P.Nozières, 'Crystal growth and crystal curvature near roughening trnasitions in hcp ^{4}He', J. Phys. **46** 1987-2007 (1985).

[199] D.J. Wollkind and L.A. Segel, 'A nonlinear stability analysis of the freezing of a dilute binary alloy', Philos. Trans. Roy. Soc. London **268**, 351-380 (1970).

[200] G.Wulff, Z. Kristallogr. Mineral. **34**, 449 (1901).

[201] E.Yokoyama and T.Kuroda, 'Pattern formation in growth of snow crystals occuring in the surface kinetic process and the diffusion process', Phys. Rev. A**41**, 2038-2049 (1990).

[202] G.W.Young, S.H.Davis and K.Brattkus, 'Anisotropic interface kinetics and tilted cells in unidirectional solidification', J. Cryst. Growth **83**, 560-571 (1987).

[203] M.Zimmerman, M.Carrard and W.Kurz, 'Rapid solidification of Al-Cu eutectic alloy by laser remelting', Acta Metall. **17**, 3305-3313 (1989).

[204] M.Zimmerman, A.Karma and M.Carrard, 'Oscillatory lamellar microstructure in off-eutectic Al-Cu alloys', Phys. Rev. B**42**, 833-837 (1990).

Index